MEGASCIENCE: THE OECD FORUM

PARTICLE PHYSICS

ORGANISATION FOR ECONOMIC CO-OPERATION AND DEVELOPMENT

ORGANISATION FOR ECONOMIC CO-OPERATION AND DEVELOPMENT

Pursuant to Article 1 of the Convention signed in Paris on 14th December 1960, and which came into force on 30th September 1961, the Organisation for Economic Co-operation and Development (OECD) shall promote policies designed:

- to achieve the highest sustainable economic growth and employment and a rising standard of living in Member countries, while maintaining financial stability, and thus to contribute to the development of the world economy;
- to contribute to sound economic expansion in Member as well as non-member countries in the process of economic development; and
- to contribute to the expansion of world trade on a multilateral, non-discriminatory basis in accordance with international obligations.

The original Member countries of the OECD are Austria, Belgium, Canada, Denmark, France, Germany, Greece, Iceland, Ireland, Italy, Luxembourg, the Netherlands, Norway, Portugal, Spain, Sweden, Switzerland, Turkey, the United Kingdom and the United States. The following countries became Members subsequently through accession at the dates indicated hereafter: Japan (28th April 1964), Finland (28th January 1969), Australia (7th June 1971), New Zealand (29th May 1973) and Mexico (18th May 1994). The Commission of the European Communities takes part in the work of the OECD (Article 13 of the OECD Convention).

Publié en français sous le titre :
PHYSIQUE DES PARTICULES

Foreword

The Megascience Forum was established on 1 June 1992 by the Council of the OECD as a subsidiary body of the Committee for Science and Technological Policy. The goal of the Forum, in the context of the work programme of the Directorate for Science, Technology and Industry, is to help ensure the exchange of information and open and substantive discussion of issues relating to existing and potential megascience projects among Member country governments and the scientific communities involved in very large scientific undertakings. In this context, expert meetings are held in disciplines that make significant use of megascience, and the results of these meetings are published in the Megascience Forum series.

The expert meeting on particle physics was the sixth in the series of such meetings and was decided upon by the Megascience Forum at its third meeting in July 1993. As agreed at the following meeting of the Forum, the boundaries of the field under review in this report include high energy nuclear physics and particle astrophysics, which are closely linked to particle physics. An expert group met at the Paul Scherrer Institute, in Villigen, Switzerland, at the invitation of the Swiss Government. The meeting, held 16-18 May 1994, was attended by 51 participants from 17 OECD Member countries and by eight observers from various international organisations concerned with particle physics. Seven observers from Hungary, India, Mexico (which became an OECD Member country in May 1994), and the Federation of Russia were also present. The goal of the meeting was to examine the situation of research in particle physics and to discuss international co-operation in this area today, and, more importantly, in the future.

This volume presents the document prepared for and the results of those discussions. Part I presents the main conclusions of the expert meeting, which were largely endorsed by the Megascience Forum. The preface to the first part, prepared by the President of the Forum, indicates the points on which the Forum diverged from the conclusions of the expert meeting. Part II, prepared by a group of six eminent particle physicists, is the comprehensive overview of scientific and organisational issues in the field of particle physics that served as the basis for discussion at the Villigen meeting: it presents the development of the scientific field, the use of existing facilities, the socio-economic benefits of particle physics, and existing mechanisms for international co-operation, before turning to new projects and long-term developments and scenarios for implementing future accelerators and detectors. Three annexes are also included: Annex 1 consists of three tables which draw together quantitative information on this area, while

Annexes 2 and 3 discuss the interface between particle physics and high energy nuclear physics and between particle physics and astrophysics, respectively. This volume thus presents various aspects of the situation in particle physics, with special attention to the issue of international co-operation, as of September 1994.

It is important to emphasise the novelty of this effort, which brought scientists and science administrators together to discuss not only their discipline but also the political context that affects the development and financing of the research that is necessary for scientific progress.

The present volume is published on the responsibility of the Secretary-General of the OECD.

Table of contents

Part I
CONCLUSIONS OF THE MEGASCIENCE FORUM ON PARTICLE PHYSICS

Part II
PERSPECTIVES ON PARTICLE PHYSICS AND INTERNATIONAL CO-OPERATION

List of Figures

7

List of Tables

Part I

CONCLUSIONS OF THE MEGASCIENCE FORUM ON PARTICLE PHYSICS

Chairman's Preface

The field of particle physics was certainly one area of concern at the meeting of the Committee for Scientific and Technological Policy (CSTP) at Ministerial level which led to the establishment of the OECD Megascience Forum in 1992. It is not surprising, then, that when the US Congress cancelled funding of the Superconducting Super Collider in October 1993, the Megascience Forum organised an expert meeting to discuss this important megascience discipline. The meeting drew more than ordinary attention from governments and national agencies involved in particle physics.

In reading the conclusions of that meeting, it is important to keep in mind the basic purpose of expert meetings. As the name implies, an expert meeting brings together eminent specialists from all over the world to address the issue of global co-operation in a given discipline. Though scientists in many fields, not least in particle physics, have grown accustomed to working together in large international facilities, the present emphasis on global co-operation and some of the particular concerns or questions it raises very much have their origin with governments. In the expert meetings, scientists are asked to discuss their experience with global co-operation, the mechanisms they have established, and the efforts they have engaged in or expect to engage in. In addition, they are asked to reflect on priorities or on the appropriateness of existing mechanisms to cope with ever tighter constraints due to increasing costs, declining budgets and a scientific community more widely dispersed over the globe than ever before. In an area as costly as particle physics, these questions impose themselves on the scientific community and on governments to an even greater degree than in most other fields of science. It is not surprising, then, that the scientific community and governments do not always share the same views on all issues.

During its 5th meeting, on 7 and 8 July 1994, the Megascience Forum discussed the Conclusions of the Expert Meeting on Particle Physics that follow. The Forum expressed great appreciation for the extensive and frank discussion of global co-operation that had taken place. It endorsed many of these conclusions. Yet it concluded that, in some areas, the perspective of governments is, inevitably, beginning to diverge from that of scientists.

A first point has to do with the scope of megascience, *i.e.* with the cost threshold above which global co-operation must be seriously envisaged. The Forum feels that in particle physics, even at lower levels of energy (below 10 GeV) or investment (under $1 billion), attempts should be made to replace regional approaches by global co-operation. The second point is related. While the Forum does not underestimate the importance of maintaining sufficient competition and diversity of scientific approaches, it is inclined to ask, with some insistence, whether we can afford competition that relies on building the same type of machines with comparable scientific programmes and potentialities in different countries. In the Forum's view, it is very important to work out practical ways to connect priority setting and planning in the scientific community with the setting of boundaries and policies, which is usually the realm of government. Therefore, some form of close interaction between the scientific community and governments, at a relatively early stage of planning and at the global level, would be appropriate.

The Forum will consider the issues further, as it prepares for the next meeting of the CSTP at Ministerial level, scheduled for September 1995.

Conclusions of the Expert Meeting

The meeting expressed its satisfaction with the way the authors of the background report, "Perspectives on Particle Physics and International Co-operation" (see Part II), had reviewed the present status and future development of particle physics. It also endorsed the description of the socio-economic context of particle physics in this report. Moreover, it pointed out that, whereas much of the present-day discussion on large facilities focuses on budgetary constraints, it should be borne in mind that embarking upon big co-operative scientific enterprises contributes in an essential way to the cultural development of mankind. This is all the more true if the collaboration in these projects is based on a better understanding of the different cultures and administrative schemes in the countries involved.

The value of the meeting certainly lay in providing science policy officials with an international perspective on the large and expensive particle physics projects being planned for, and in discussing the mechanisms within the scientific community for further extending global priority setting and co-operation. When and how these processes within the scientific community should be linked to consultation and negotiation mechanisms between governments was discussed, but integrating the different perspectives of the two sides into an effective decision-making scheme clearly needs further study. So do the setting of priorities and the linking of decisions across different fields of science.

1. The challenges ahead

There is widespread agreement on the scientific challenges in high energy physics/ particle physics: the further testing and exploration of the Standard Model; searching for the physics beyond this Model, where the answers to some major questions must lie; and the investigation of the phenomena that controlled the evolution of the very early Universe.

Equally widespread is the conviction that there is a very short list of future accelerators that should be considered as having top priority for probing these areas. In the megascience range (construction costs of the order of $1 billion or more), the list includes, as a first priority, a high energy hadron collider [for which the Large Hadron Collider (LHC) is now the only immediate candidate], and, as a next step, a large electron-positron linear collider. In the sub-megascience range (construction costs of the order of several hundred million dollars), high luminosity facilities, especially B factories, feature prominently. Two B factories have been approved; so far, in view of their more limited size, such facilities have been dealt with individually, on a national or regional basis. A list of sub-megascience facilities might also include the Large Electron Positron Collider upgrade (LEP-II, CERN), Fermilab's upgraded injector, and the Russian UNK-600 (Protvino) proton synchrotron. Some aspects of astrophysics touch on particle physics, and that community should therefore be consulted at an appropriate stage when future very large facilities are being considered. A similar case was made for the high energy range of nuclear physics, but it received less support. In particular, key questions in the emerging field of astroparticle physics very likely can only be addressed by large underwater or underground non-accelerator facilities whose scale brings them to the threshold of megascience. Focusing on one or two international detectors of sufficient size was considered to be important for ensuring orderly progress. In all these cases, global co-operation should be recognised as the only way to realise the opportunities ahead.

Interregional collaboration in exploiting major regional facilities in North America, Europe, and the Far East remains a cornerstone for progress in particle physics. Russia remains an important player as well. Global use of these facilities is increasing, both in the Americas and in Asia. However, concern was expressed about the situation in Eastern Europe.

2. The Large Hadron Collider

The meeting strongly endorsed early approval of the LHC. Financial participation beyond the commitments of the members of the European Laboratory for Particle Physics (CERN), by regions or communities outside Europe, was seen as the best way to optimise scientific and technical input and exploitation of the machine. This first opportunity for intensive global co-operation in the construction, as well as the utilisation, of a particle physics accelerator should be seized. CERN and the governments concerned were called upon to elaborate the terms of participation as quickly as possible, balancing the importance of wide utilisation and involvement by different regions in the construction, bearing in mind that this case will set a precedent for future global co-operative projects.

3. Electron-positron linear colliders

Linear colliders present a future possibility for international collaboration of global interest. It is encouraging that collaborative R&D efforts for this category of accelerators are now formalised through a Memorandum of Understanding (MOU) among the institutions concerned. All relevant parties (institutes, governments) should now do their utmost to work out methods of maintaining, first, free and healthy competition, then, in later stages, global co-operation; they should pay particular attention to the moment at which governments need to become collectively involved. In order not to forego opportunities for global collaboration, it is important to recognise early on the need to take a decision to continue with a scheme of global co-operation.

4. Consultation and co-operation mechanisms in the scientific community

Appreciating the extent to which international consultation and co-operation at the scientific level takes place thanks to bodies such as the European Committee for Future Accelerators (ECFA) and the International Committee for Future Accelerators (ICFA), the experts discussed several aspects of scientific foresight, priority setting and forward planning of facilities that should be more explicitly addressed in the future:

- In the light of some recent examples, it is important that maximum international collaboration should be sought at a sufficiently early stage of project development.
- The need to maintain sufficient diversity in approach and scale requires careful consideration of proposals both for new facilities and for closing or refurbishing old ones. Construction of new facilities at existing sites was judged to be cost-effective and efficient. Also, the scientific and technological capabilities, modernity, and running costs of all facilities available globally, including those in the New Independent States (NIS), should be taken into account when considering new ones.
- It is perfectly natural to distinguish between national and international facilities on the basis of scale, though there may be other reasons to choose the national or the international route. If international co-operation is to be maximised, and if the best use is to be made of financial and human resources, national or regional decisions will be influenced by international considerations. Indeed, a balance between national and international investments is important, and co-ordinating decisions on national or regional facilities will be of great benefit as well.
- In addition to the scientific objectives themselves, important elements in the scientific planning of new facilities have been: the need to serve a growing scientific community, on average with a constant or declining budget, and the need for competition to ensure that there is adequate and independent verification of the scientific robustness of results. Do these elements continue to be appropriate? Should other aspects be considered?

ICFA ought to consider whether these issues should lead to a modification of its composition and mode of operation. For maximum credibility and authority, it is important that high energy and particle physicists from countries other than the current base regions are adequately represented.

ICFA has described four organisational models for large facilities. In determining which one best applies to a certain facility, the advantages and disadvantages of the various models should be explicitly considered. They are: *a)* national or regional facilities, built and operated by the host country or region; *b)* national or regional facilities in which a host country or region plays a dominant role, with a sizeable contribution to construction from others (*e.g.* Hadron Elektron Ring Anlage, HERA); *c)* very large facilities to be planned, constructed and operated under the joint responsibility of several countries or laboratories participating with comparable shares; and d) very large projects to be set up as an international organisation (*e.g.* CERN).

It was not considered appropriate to express clear opinions on models for interregional facilities, but it was noted that an interregional CERN model was only one possibility and may not be the first one to consider. For example, the European Synchrotron Radiation Facility (ESRF) model, with a different type of agreement among governments and with a legal status of private company, was pointed out as another possibility.

As to astroparticle physics, it would be desirable to form an appropriate international representative body to provide an interface between this community and government policy makers, perhaps with International Union of Pure and Applied Physics (IUPAP) sponsorship.

5.   Priority setting in the scientific community

It was believed that the major priority decisions in particle physics for the foreseeable future are few in number, and that, in general, they have been dealt with adequately. However, current mechanisms may not take sufficient account of the importance of projects at the interface with neighbouring fields. As to duplication of facilities, it was stressed that the need for scientific verification of results, diversity of methods, and competition have been and will remain important considerations in the planning of facilities. It was clear, however, that the increase in the scale of facilities may well reach the point where desirability clashes with declining budgets. Other solutions will then have to be found.

Assigning priority to a particular type of particle physics facility does not mean, of course, that no further technical or other choices need to be made; these may, in turn, affect the priorities as originally set. Moreover, in the future, the priority-setting process may be broadened to include a wider range of countries or regions. This concerns both the ever smaller number of very large facilities and the larger number of medium-sized facilities, which might be distributed more widely over the world. This evolution may also have an impact on recent or ongoing priority decisions.

6.   Boundary conditions

It is to be expected that governments will increasingly and more forcefully express the need for global co-operation. If nothing else, budgetary constraints will cause them to do so. This will put pressure on the current priority-setting mechanisms in the scientific community and require them to address priorities on a global scale more often and at earlier stages. For planning to be realistic, the scientific community and the science agencies should receive predictable and, if possible, stable information about the medium- and longer-term budgetary envelopes. Stable and reliable international partner-

16

ships require, to an even greater degree, long-term financial commitments, with due regard to different national budgetary procedures.

At the same time, the foregoing points to the need for appropriate forms of interaction between governments or agencies and the scientific community at the stage, neither too early nor too late, where development and planning of new facilities result in plans becoming fairly fixed.

7.   Conditions of access to large facilities

The operation of a large accelerator with a few big detectors for scientific experiments differs considerably from that of other facilities where many groups run fairly short experiments by setting up instruments at many beam sites. This should be borne in mind when access arrangements are considered. For one thing, it has led to the formation of the very large teams under whose umbrella researchers work.

The scientific world has benefited enormously from the fact that the results of basic science are published and available to all. Once the rules for financing the construction and for operating a given facility have been established, the selection of scientific projects should be based only on the scientific merit of individual proposals. The principle of ensuring open access as much as possible is central to the ICFA guidelines. Host institutes benefit as much from their users' intellectual and technical capabilities as the latter benefit from the former's facilities. The spirit of the ICFA guidelines was endorsed, although it was pointed out that some aspects might need reconsideration now that the number and geographical distribution of sites is quite different from what it was in the past.

The attention of governments is drawn to the enduring difficulty of obtaining visas for scientists who have to perform experiments at large facilities abroad. These barriers should be removed for the scientists and, when long stays are involved, for their families as well.

8.   Governments and consultation

Nationally, and even regionally, there is a reasonable degree of exchange of information and consultation among the particle physics community, national research councils/organisations, and governments. The question is when interregional consultation among governments is appropriate and in what form.

It was considered useful to distinguish three stages. In the first phase, where various possible projects are being investigated in the scientific community in order to come to a first assessment of scientific value and technical feasibility, the emphasis should be on the exchange of information. If ICFA encourages representatives from governments to be involved in its discussions from time to time, this could be a way to help inform governments.

17

Once the scientific community is about to set its priorities, closer communication between national research councils/organisations as well as between governments seems appropriate. The perceptions, financial limits, and policy contexts of the group of governments concerned, on the one hand, and scientific considerations, on the other, should be integrated in the final decision process.

The third stage requires communication among governments and is, therefore, to a large degree, parallel to and synchronous with the second. Various ways to ensure interregional communication among governments and between governments and the scientific community have been discussed. It is clear that the question of when and how communication should be organised at a regional and interregional level among governments themselves, and between governments and the scientific community, needs further consideration and discussion. This matter will, of course, be taken up by the Megascience Forum.

Two important elements were stressed:

- First, appropriate means of organising discussions among the three main regions involved in particle physics, *i.e.* North America, Europe, and Asia, should be investigated. Different degrees of autonomy in each of these regions and widely varying political systems make this a difficult but important task.
- Second, assessing and assigning priorities to facilities in different fields of science have to be dealt with by science agencies and governments. When new facilities are planned, others must be considered for closure, a practice already well-established in particle physics. The idea of negotiations concerning a "basket of facilities" (or "multiple megaproject approach" as it is called in Part II), *i.e.* simultaneous negotiations on a number of more or less similar megafacilities in different scientific disciplines, was considered unrealistic and met with considerable scepticism.

9. Global, not just first-world, co-operation

The progress of science and the strengthening of national scientific, technological, and industrial capabilities point to the importance of extending the current North America-Europe-Japan base of global co-operation to Russia, to the other Central and Eastern European countries and to those of the Asia-Pacific region, in particular China, and to other developing countries. The preservation of the scientific potential of Russia and other members of the NIS and the maximum exploitation of their facilities and expertise are extremely important if the human and physical capital built up there over decades is not to be wasted. Megascience projects should not be monopolised by the developed countries. For developing countries, being involved in appropriate ways in the great scientific enterprises of mankind is of great cultural, psychological, and practical value. It requires, however, from the scientists concerned, a close connection with, or a sense of responsibility for, the training of new generations of their societies' leaders and the development of industry.

18

Since astroparticle physics has been explored by scientists from developing countries, among others, anticipated megaprojects in this field should naturally maintain this tradition.

The experts noted with interest that the UNESCO Physics Action Council (PAC) could, in principle, play a substantial role through its Large Physics Facilities Working Group, especially with regard to developing countries. Also, concern to associate the developing countries would certainly warrant ensuring conditions for access to facilities for their scientists. The accelerator laboratories, as well as ICFA, would do well to look further into these matters. The authors of the background report suggested forming regional centres, and this certainly deserves wider attention and follow-up by bodies such as the UNESCO Committee, IUPAP, and ICFA.

Part II

**PERSPECTIVES ON PARTICLE PHYSICS
AND INTERNATIONAL CO-OPERATION**

Foreword

This report, an overview of issues in particle physics, is the collective work of six particle physicists deeply involved in the field: M.V. Danilov, Institute for Theoretical and Experimental Physics (ITEP), Moscow; A. Donnachie, University of Manchester; G. Flügge, Reinische Westphälische Technische Hochschule, Aachen; W. Willis, University of Columbia, New York; S. Wojcicki, Stanford University, California; S. Yamada, University of Tokyo. P. Petiau, of the École polytechnique, Paris, was of invaluable assistance to the OECD Secretariat in the preparation of the report. The subjects covered in the different chapters are:

- a general overview of the scientific objectives of particle physics;
- an analysis of the use of existing facilities, indicating their scientific impact and potential and the level of their budgets;
- a survey of the socio-economic benefits derived from the past achievements of particle physics;
- an investigation of the existing mechanisms of international co-operation, with their advantages and drawbacks;
- a presentation of projects that are new, recently approved, or still under consideration, and of longer-term developments;
- an attempt to invent scenarios for implementing future megaprojects.

Each chapter was prepared by several experts and was discussed and endorsed by the entire group.

Some limits have to be set – arbitrarily to some extent – for defining installations considered as megascience, *i.e.* those whose high cost and scientific impact justify international co-operation for their construction or for their optimal scientific exploitation. The selection criterion used in this review is based on the energy of the beam(s) circulating in the machine: the only facilities taken into consideration are accelerators with beam energy exceeding 10 GeV, or colliders providing an energy of at least 10 GeV in the centre-of-mass reference frame of the colliding particles.

Installations which do not include accelerators have not been considered, although some might give important scientific results, mainly because so far no project of this kind has needed the level of funding required by large accelerators.

The laboratories, institutes and centres mentioned in the text are often designated by their acronym. A list of acronyms can be found at the end of the volume.

Executive Summary

After a short overview of the scientific goals and past achievements of particle physics and a survey of the use of existing installations, this report describes future projects in the field and investigates the new opportunities for international co-operation that these projects present.

1. The development of particle physics

As particle physicists have been able to access higher and higher energies and probe the structure of matter to smaller and smaller distances, layer after layer of complexity has been removed, revealing a unifying simplicity. This has culminated in the "Standard Model" of particle physics; this model successfully describes all phenomena at presently available energies in terms of a limited number of elementary particles which interact through the operation of a few fundamental forces. The creation of the Standard Model must rank as one of the great scientific achievements of this century.

In studying the elementary particles and their interactions, physicists are probing backwards in time, back to the earliest moments of the Universe, back beyond the limit accessible to optical and radio astronomy. Underlying their attempts to uncover the ultimate structure of matter is the even more ambitious goal of describing how the early Universe evolved from the cosmic fireball of the Big Bang.

Despite the astonishing success of the Standard Model, it cannot be the whole story. There are many questions which this Model is unable to answer, ranging from why the magnitude of the electric charge on the proton is the same as that on the electron (when they appear to be entirely different entities) to why we live in a matter-dominated Universe (when the Big Bang would produce matter and antimatter in equal quantities). There are strong reasons to believe that many of the outstanding questions will be answered by the next generation of particle accelerators.

2. Use of existing facilities in particle physics

The currently active facilities, working either at the high energy frontier or as "factories", *i.e.* with the highest possible beam intensity, are located in a relatively small

24

number of laboratories (less than ten) with large accelerator complexes and which generally support broad research programmes. Depending on the type of particle collisions, they can be classified in four categories: electron-positron colliders, electron-proton colliders, proton-antiproton colliders, and fixed-target proton accelerators. Facilities in different categories generally carry out different scientific programmes, but they are frequently able to address the same scientific question in complementary ways.

In order to carry out the most effective research, co-operation by a number of user groups has been the rule for many decades; due to the complexity and the cost of the experimental installations, the work of constructing detectors and analysing the data is usually shared by the user teams. To build and operate the huge detectors now installed at the most attractive facilities, giant collaborations, in which several hundreds of physicists join efforts, have been formed. The corresponding costs, which will reach levels of several hundred million dollars for LHC experiments, are mainly underwritten by the funding agencies of the laboratories involved in the collaborations.

Specific management and review processes have been set up to guarantee that the focus on the final physics aims is firmly maintained, while ensuring technical flexibility and financial control.

Past experience shows that the scientific community has developed an impressive capacity to make the most efficient use of former investments by reconverting facilities no longer at the forefront of research to new applications: the 28 GeV CERN Proton Synchrotron (PS) or the Brookhaven 30 GeV Alternating Gradient Synchrotron (AGS) are striking examples.

In order to obtain the best scientific payoff from the available resources, older facilities still producing valuable results have nevertheless had to reduce their activity, or be closed, in order to make possible the construction of new instruments for a more attractive scientific programme. In the history of particle physics, difficult decisions of this sort have had to be taken.

A careful examination of the scientific programme of the present largest facilities shows their high degree of complementarity: most are unique in their category or in their energy range. Duplication in the experimental set-ups is generally limited to the minimum needed to confirm the exactness of the results.

3. Socio-economic benefits of particle physics

The main reasons for high energy physics research are cultural; the resulting benefits are somewhat intangible and difficult to quantify. Certainly no rigorous cost/benefit analysis is possible. Research into the fundamental workings of the Universe, such as that performed in particle physics, enriches our culture and helps to stimulate other scientific fields. Perhaps its largest immediate impact is in the area of education: the questions the field attempts to answer, the sophisticated instrumentation it uses, and the exciting results it provides all act as important stimuli in attracting young people into technically oriented education. A better-educated society is the end result.

But these primary reasons for studying particle physics should not draw attention away from the wide range of secondary benefits derived from the use, in other areas, of tools and techniques developed for the study of the fundamental nature of matter. Other scientific disciplines, medicine, computing, and the electronics industry are all much richer today because of the earlier investment in particle research. On the basis of past history, there is every reason to expect that this trend will continue, even though the exact scenario cannot be predicted today.

4. Existing mechanisms for international co-operation

International co-operation is a long-standing tradition in high energy physics. In terms of scientific exchanges and sharing of facilities, the field is very open. International co-operation often leads to the optimal selection of research goals. In order to construct and operate large facilities and/or experiments, it brings together intellectual, technical, material, and financial resources from different countries. Through international co-operation, particle physics has been very successful in generating intellectual achievements, promoting international understanding, and setting models for collaboration in other scientific areas.

International collaboration may have various goals:

a) Exchange of scientific information through the organisation of conferences and symposia or exchanges of scientists between different laboratories. In this domain, Commission 11 (C11) of IUPAP (the International Union of Pure and Applied Physics), and the relevant sections of national or regional physical societies play a major part. Many bilateral government agreements or programmes foster exchanges of scientists.

b) Co-operation for the construction of large facilities. In this case, agreements are necessary at the government level. Up to now, such co-operation has only been implemented for facilities in Europe. CERN is an international European organisation whose Convention has been signed by 19 member countries. For the construction of HERA, bilateral agreements were concluded between Germany and several countries, both in and outside Europe. Other models have been developed in Europe in other fields of science (neutron sources, synchrotron radiation installations, etc.).

c) Collaborations to participate in an experiment at a given facility. It is common practice for most experiments at large facilities to be performed in the framework of international co-operation, whether the facility is national or international. The procedures have been clearly defined by the scientific community, and formal agreements are generally prepared at the level of the funding institutions of the scientific teams.

There are two bodies within the high energy physics community whose aim is the promotion of international co-operation. In Europe, the European Committee for Future Accelerators (ECFA) has been established to help co-ordinate European activities by providing advice from the scientific community to the management of CERN, to DESY, and to funding agencies in European countries and by organising seminars. At the world level, the International Committee for Future Accelerators (ICFA) was set up by IUPAP-C11 to study the issue of the very large accelerators that might be built and used in world-wide co-operation. While discussing the technology and the plans for such machines, ICFA established guidelines for the international use of major regional facilities which are respected by all laboratories in the field.

ICFA serves to promote communication and understanding among the world's high energy physicists. It still has some way to go to achieve its original aim, and it might be useful to reinforce the organisation. One possibility might be to strengthen ICFA's link to the user community, perhaps followed by an expansion of its role with respect to governments.

5. New projects and long-term developments

In the past two decades, colliding beam experiments have been the most fruitful ones in particle physics. It is therefore not surprising that all plans for future projects in this field are of this collider (storage ring) type. Progress is made either by increasing the energy of the colliding particles significantly or by increasing the intensity of the beams in so-called "factories".

If discussion is restricted to colliders with a centre-of-mass energy of more than 10 GeV, all medium-energy high-intensity colliders are of the B factory type. Two of these have recently been approved for construction at SLAC (United States) and at KEK (Japan).

At the energy frontier, two routes are being explored: proton-proton (pp) and electron-positron (e^+e^-) collisions. Consistency arguments within the framework of the Standard Model indicate that at an energy of a few TeV, new phenomena should appear. Since protons are aggregates of quarks and gluons, each carrying only a fraction of the total momentum, the energy necessary for pp collisions would exceed 10 TeV. The only project at this energy ready for approval in the near future is the Large Hadron Collider (LHC) at CERN. It is expected to operate at an unprecedented energy (14 TeV centre of mass) and beam intensity. This puts very stringent demands on the technology of the machine, in particular the superconducting magnets and the detectors. According to present plans, physics experiments could begin in 2003, provided sufficient additional resources (some 20 per cent of the total cost of SF 2.5 billion) can be made available for the project. Otherwise the completion of the project would be delayed by up to two years.

Electron-positron collisions at a few TeV would present an equally powerful tool for studying the energy frontier. In many respects, they would be complementary to pp collisions. The next generation of e^+e^- colliders, which would exceed the energy of LEP-II (about 190 GeV), have to be of the linear collider type. This new technique, pioneered at SLAC, still requires significant R&D work before a definitive proposal can be formulated. A world-wide collaboration for optimising and testing the required techniques has been established. It will be a few years yet before the best solutions emerge and their feasibility is demonstrated. Current efforts concentrate on a machine with an initial centre-of-mass energy of about 500 GeV; it would provide an excellent tool for studying the top quark and provide information complementary to that provided by the LHC. One of the goals of the design is to ensure the later extension of the energy to 1 to 2 TeV.

6. Scenarios for implementing future accelerators and detectors

Effective mechanisms have already been developed in particle physics for comparing the scientific merits and cost effectiveness of different projects. They include extensive scientific discussions and consultations among a wide variety of interested parties: communities of theorists and experimentalists, accelerator physicists and engineers, national and international advisory panels, physical societies, international bodies for particle physics (ECFA, ICFA), and funding agencies.

These mechanisms, together with some historical developments, have made it possible to formulate a well-grounded proposal for a future particle physics programme. It includes upgrades of existing facilities and construction of new accelerators. Projects for several new facilities are now at different stages: two B factories have been approved (in the United States and Japan), and the LHC is on the verge of receiving approval from the CERN Council.[1] Very extensive R&D work on a future electron-positron linear collider (EPLC) is underway.

B factories are being built as national projects, although the definition of the projects and the choice of parameters were based on broad international discussions. Active negotiations are taking place on possible Chinese and Russian participation in the construction of these facilities.

The LHC is planned as a new large facility at CERN. Contributions from non-member countries would allow the project to proceed more rapidly and improve its scientific programme. Several countries, including Canada, China, India, Japan, Russia, and the United States, are considering a possible contribution to the construction of the LHC. Participation of non-member countries in the construction of the LHC can be organised using the existing practice of co-operation agreements. Closer institutional relations might be appropriate in the case of large contributions. For example, an ''associate status'' might be introduced to allow a country offering a sufficiently large cash contribution to participate in CERN's scientific policy discussions, have representation in CERN official bodies, and perhaps obtain additional benefits. This concept is still being developed and seems the appropriate way forward, as the possibility of non-European countries becoming CERN members looks too remote to be discussed in the context of the LHC. Approval of the LHC in 1994 is extremely important for maintaining momentum and for simplifying the negotiations on contributions from non-member countries.

Vigorous research and development work on the EPLC is now being done, to a large extent in the framework of international collaboration. This complies with the ICFA recommendations for establishing international collaboration on large accelerators at the

---

1. The CERN Council approved the LHC in December 1994.

28

R&D stage and with its further recommendation that the scientific community should formulate a relatively precise proposal, including realistic cost estimates and site criteria, before formally approaching governments.

There are reasons to believe that the EPLC might be constructed through international co-operation. However, the selection of the site and the technology presents serious problems that may hinder the definition of a common project. The organisational structure is still a completely open question and will depend on the actual level of the cost of the machine, on the number of scientists needed for optimal realisation of the physics programme, and on the specific motivation of a government prepared to pay a higher share as "site premium". Lessons might be drawn from the ESRF (European Synchrotron Radiation Facility) for the process of approval as well as for the legal status.

7. Conclusion

Particle physics has been totally open to international co-operation for the last 50 years at least: unrestricted publication and diffusion of scientific results and relevant technologies, exchange of physicists, and extensive sharing of the use of existing facilities have been the rule.

As a result, the field has successfully generated some outstanding intellectual achievements, of which the elaboration of the Standard Model and its experimental verifications offer the most striking example. It has provided society with cultural, educational, and economic enrichment. An important side benefit has been the promotion of better international understanding and the setting of models for such co-operation in other fields.

During the past decades, various modes of international co-operation have been developed, depending on specific circumstances. The success of CERN is universally acknowledged.

Co-operation in particle physics has been initiated and continuously nourished by scientists themselves, the leading players invariably being physicists or engineers, and the basic motivation has always been the best use of available facilities and resources for the progress of science.

The recent changes in the world economic situation, competition with other fields of science, and the growing size and cost of new facilities dictate some enlargement of the scope of this co-operation and conceivably the establishment of new types of agreements, extrapolated from existing models inside or outside the field. When devising these new agreements, it is essential not to sacrifice optimal scientific or technical solutions for reasons of political expediency and to make the best use of past experience, both successes and failures.

Definitions and Units in Particle Physics

Energy

The relevant scale of energy in particle physics ranges from the MeV (one million electron-Volts) to the GeV (one billion electron-Volts = 1.6 10^{-10} Joule) and to the TeV (1 000 Gev). As an example, each Large Hadron Collider (LHC) beam (3×10^{14} protons of E = 7 TeV) will carry an energy of 330 MegaJoule.

Mass

According to Einstein's relation, $E = Mc^2$, particle masses M are commonly indicated using the value of the product Mc^2 measured in MeV or GeV (1 GeV/c^2 = 1.8 10^{-27} kg); c denotes the speed of light. Examples include:

– electron mass: Mc^2 = 0.51 MeV;
– proton mass: Mc^2 = 0.938 GeV;
– Z^0 boson mass: Mc^2 = 91.2 GeV.

Luminosity and cross-section

When two beams of particles interact inside a collider, the number N of collisions occurring every second is given by:

$$N = L\sigma$$

a) The "luminosity" L is a parameter of the collider directly proportional to the beam intensities, and inversely proportional to the beam size at the interaction point. L is measured in cm^{-2} s^{-1}. Typical or expected values range from 10^{30} to 10^{34} cm^{-2} s^{-1}.

b) The "cross section" σ is a physical parameter, measured in cm^2, which characterises the probability of the individual collisions, or of any specific process occurring during the collision (production of a particular final state). Typical orders of magnitude are widely spread, depending on the elementary process and on the energy:

– 10^{-25} cm^2 for proton-proton collisions;

– 10^{-32} cm^2 for peak production of Z^0 at LEP;

– 10^{-38} cm^2 for Higgs boson production at LHC, with subsequent decays into two Z^0 bosons, then into four leptons.

Chapter 1

The Development of Particle Physics

1. Introduction

For nearly a century, physicists have sought to uncover the ultimate structure of matter, the fundamental particles from which the Universe is built, and to establish the laws that bind these constituents together. Particle physicists now believe that Nature's remarkable diversity and richness stem from a small number of principles of compelling simplicity, and that these principles manifest themselves through the operation of a few fundamental forces on a limited number of elementary particles. This discovery must rank as one of the significant scientific achievements of this century. In studying these particles and their interactions, physicists are, at the same time, probing backwards in time, back to the earliest moments of the Universe. Underlying their attempts to reveal the relationships between matter and energy is the even more ambitious goal of describing how the early Universe evolved from the cosmic fireball of the Big Bang. One of the most exciting developments in recent years has been the growing symbiosis between elementary particle physics and cosmology – the particle physicists attempting to reproduce the conditions of the very early Universe under controlled conditions in the laboratory, the astronomers looking far out into space and trying to decipher the clues which this gives to the state of the very early Universe.

Throughout the years, layer after confusing layer has been peeled back as particle physicists have been able to reach ever higher energies and probe ever smaller distances. As each layer has been removed, apparent complexity has been replaced by a unifying simplicity. To reach beyond the present layer will require new and diverse tools: higher energies to access foreseen thresholds and to probe structure with ever greater resolution; specialised machines, not necessarily of particularly high energy, to tackle very specific problems that are only amenable to this approach; and non-accelerator experiments, such as those, deep underground, that search for rare occurrences, such as proton decay or anomalous behaviour in the neutrinos from the sun or supernovae.

This chapter summarises briefly the development of the subject from the late nineteenth century to the present day (see Box 1), and outlines the questions to be answered by the next generation of particle accelerators.

Box 1. A Brief History of Particle Physics

1808	Dalton proposes the modern atomic theory of matter
1869	Mendeleev's periodic table of the elements
1896	Becquerel observes nuclear radiation
1897	Thomson discovers the electron
1911	Discovery of the atomic nucleus
1913	Bohr's model of the atom
1923	Compton scattering
1924	de Broglie conjectures wave-particle duality
1927	Davisson and Germer verify wave-particle duality Schrödinger's wave equation for particles
1928	Dirac's equation and the prediction of antimatter
1931	Pauli postulates the neutrino
1932	Anderson discovers the positron Chadwick discovers the neutron
1933	Fermi's theory of weak interactions
1935	Yukawa predicts the pion
1937	Anderson and Neddermeyer and, independently, Street and Stevenson discover the muon in cosmic rays
1938	Klein suggests an "intermediate vector boson" (the W) as a carrier of the weak force
1947	Powell *et al.* discover the pion in cosmic rays
1948	Pion produced in an accelerator experiment at Berkeley Rochester and Butler discover strange particles in cosmic rays
1953	Gell-Mann, Nakano and Nishijima propose strangeness
1956	Lee and Yang suggest that mirror symmetry may be violated in weak interactions Wu and Ambler discover violation of mirror symmetry in weak interactions Reines and Cowan observe neutrino interactions
1957	Schwinger's attempt to build a unified model of weak and electromagnetic forces
1960-61	Gell-Mann and Ne'eman suggest the "Eightfold Way" and predict Ω^-
1961	Glashow solves Schwinger's problems
1963	Ω^- discovered at Brookhaven and independently at CERN

(continued on next page)

(continued)

1964	Gell-Mann and Zweig propose quarks as hadron substructure
	Greenberg introduces colour concept
	Han and Nambu start QCD
	The Higgs mechanism invented
	Bjorken and Glashow propose charm quark
	Christenson, Cronin, Fitch, and Turlay discover CP violation
1968	Experimental proof at SLAC that the proton contains charged quarks, plus neutral objects (gluons)
	Full formulation of the Glashow-Weinberg-Salam model of the weak interaction
1970	First direct experimental evidence for colour at Frascati
	Glashow, Iliopoulos, and Maiani develop charm-quark concept
1971	't Hooft confers full theoretical respectability on the G-W-S model
1973	Neutral currents discovered at CERN
1974	Charm discovered at Brookhaven and SLAC
1975	Evidence for tau lepton at SLAC
1976	Confirmation of tau lepton at DESY
1977	Discovery of b quark at Fermilab
1979	Existence of gluon established at DESY
1983	Discovery of W^+, W^- and Z^0 at CERN
1989	Experimental proof of three families of quarks and leptons at CERN and SLAC
1994	Experimental evidence for the top quark at Fermilab

Discovery of the first family, quarks and leptons, began with the discovery of the electron by Thomson in 1897 and concluded between 1964 (with the proposal of quarks by Gell-Mann and Zweig) and 1968 (with the experimental proof of their existence at SLAC). That of the second family began with Anderson's "accidental" discovery of the muon (he was looking for the pion) and was completed in 1974 with the discovery of charm at Brookhaven and SLAC. Discovery of the third family began in 1975 with evidence for the tau lepton at SLAC and experimental evidence has just been presented for the top quark at Fermilab.

2. Photons

In 1900, Planck revolutionised physical thought when he attempted to explain the interaction of light with matter, for example the emission of light by a hot filament. Using the established classical theories of thermodynamics and electromagnetism, Rayleigh and Jeans (1900, 1909) derived a formula for the power spectrum of the electromagnetic radiation emitted when a body is heated. However, when their formula was used to calculate the total power radiated, by integrating over all the possible wavelengths of the emitted radiation, a nonsense answer was obtained: the total radiated power was predicted to be infinite.

Planck realised that the only way to avoid this result was to assume that there is a minimum amount of energy associated with any given wavelength of the radiation – a quantum of energy – and that light can be emitted or absorbed by matter only in multiples of the quantum. The minimum amount of energy, E, allowed at a given frequency ν, is given by Planck's formula $E = h\nu$, where h is Planck's quantum constant with dimensions of energy per frequency, with the minute value of 6.625×10^{-34} Joule.second. Because this constant is so small, quantum effects are easily detectable only at the atomic and sub-atomic scales.

The next major step in quantum theory was taken by Einstein in 1905, when he generalised Planck's idea to explain the photoelectric effect, *i.e.* the emission of electrons when light is shone on a metal surface. The effect had been discovered by Lennard in 1902, and it exhibited two features which could not be explained by classical physics. They are that the energy of the electrons emitted depends only on the frequency of the light, and not on the intensity, and the number of electrons emitted depends only on the intensity, and not on the frequency. Einstein proposed that all light exists in quanta, *i.e.* it comes in little packets of energy, which we now call photons, and that the energy associated with each photon is given by Planck's formula. This picture exactly fitted the experimental facts.

A specific illustration of this dual nature of light, *i.e.* that it can appear as a discrete particle (a photon) despite its undoubted wave nature, is provided by the Compton effect, discovered in 1923. Compton found that when monochromatic X-rays were scattered from a target, the scattered radiation contains two components, one at the original wavelength and one at a longer wavelength, and the larger the angle of scatter the bigger the change in wavelength. This was inexplicable in terms of classical physics, so Compton proposed treating the problem as an elastic collision between a photon and an electron in an atom in the target. In such a collision, the target particle (the electron) recoils, and so the scattered particle (the photon) has less energy. By Planck's formula, this means a lower frequency and hence longer wavelength.

3. Atoms

In the late nineteenth century, 92 atomic elements were thought to be the elementary building blocks of nature and to be indivisible. The modern atomic theory of matter was proposed by Dalton in 1808, and the periodic table of the elements was constructed by Mendeleev in 1869 (and incidentally predicted correctly three elements that had not then been discovered). Then, in the late 1890s, Thomson was forced to conclude that electrons originated from within these supposedly indivisible atoms, and by 1909 Rutherford, Geiger and Marsden had demonstrated that almost all the mass of the atom is concentrated in a small nucleus at the middle, with the electrons orbiting the nucleus at some distance.

In 1913, Bohr proposed a new theory based on the quantum ideas of Planck and Einstein. It was the simplest possible application of these ideas: just as Planck had hypothesised that light can only be emitted or absorbed in discrete quanta, so Bohr proposed that atoms can exist only in discrete "quantum states", separated from each other by finite energy differences. A simple way to think of these quantum states is as a set of orbits allotted to the electrons around the nucleus, the space between the orbits being forbidden to the electrons. With Bohr's model of the atom, physicists were able to calculate in great detail many of the spectroscopic results obtained by the experimentalists of previous decades, and to provide an explanation of many of the features of Mendeleev's periodic table of elements.

The next major conceptual breakthrough came in 1923 when de Broglie suggested that just as light waves could act like particles in certain circumstances (the discrete quanta of Planck and Einstein), so too particles could manifest a wavelike behaviour, the wavelength of the particle wave being inversely proportional to the momentum of the particle. The conjecture that electrons could behave like waves, undergoing diffraction and interference phenomena just like light or water waves, was verified experimentally by Davisson and Germer in 1927. Following on directly from de Broglie's ideas, Schrodinger formulated the equation that now bears his name and created a new dynamic theory of mechanics that transcended Newton's.

4. Nuclei

By the early 1930s the structure of the atom was well understood. The rules governing the distribution of the electrons among the atomic energy levels had been successfully worked out, and the regularities in chemical properties that Mendeleev had noticed 60 years earlier had been explained. However the nucleus still remained an enigma, with little known about it beyond the fact that it is very dense and positively charged. Experiments proved that nuclei were not featureless dense balls of positive charge but had a detailed internal structure of unknown form. The simplest atom of all is hydrogen, containing just one electron and a proton as the nucleus. As protons had been found to be ejected from more massive atoms, Rutherford suggested that the positive charge of all nuclei is carried by protons, and as the protons would not be sufficient by themselves to provide all the atomic mass, he guessed that there must be another particle as heavy as a proton but with no electrical charge: a "neutron". The neutron was discovered by Chadwick in 1932.

The force holding the nucleus together has to be very different from the electromagnetic force that holds atoms together by the mutual attraction of the positively charged nucleus and the negatively charged electrons. The nuclear force has to be independent of charge, because of the neutrons; it has to be of very short range, as a nucleus has typically a radius of 10^{-15} m compared to atomic radii of 10^{-10} m; and it has to be very strongly attractive to overcome the intense mutual electromagnetic repulsion felt by the protons when confined in such a small volume. In deference to its strength relative to electromagnetism it was named the "strong" force, although modern ideas suggest that this is something of a misnomer, as it appears to be the result of a more powerful force acting on the constituents of the nucleons.

The finite range of the nuclear force as compared to the potentially infinite range of electromagnetism implies that the carrier of the nuclear force must have a mass (as against the zero mass of the photon, the carrier of the electromagnetic force). Yukawa estimated this mass to be about 200 times the mass of the electron, *i.e.* about 1/10 of a proton mass, and the discovery in 1947, by Lattes, Muirhead, Occhialini and Powell, of the pi meson (or pion), with a mass of 140 MeV/c^2, confirmed it as the carrier of the nuclear force.

5. Neutrinos

Some nuclei are unstable and decay spontaneously. Becquerel discovered three types of radiation in 1896: alpha particles, which are helium nuclei; gamma rays, which are extremely energetic photons; and beta radiation, which consists of electrons. Beta decays, which are due to the weak interaction, exhibited two anomalous features. Unlike the alpha particles and gamma (γ) rays, which are emitted with a specific energy (as expected by energy conservation), the electrons in beta decay were emitted with a continuous range of energies: some energy was missing, as if some unseen third particle was also being produced in the decay. The existence of this particle, which carries no electrical charge and is very light (it may even be massless), was postulated by Pauli in 1931 and named the neutrino (or ν). As the neutrinos could not be detected, it was obvious that their interaction was very weak. Indeed, Pauli was so convinced of their invisibility that he wagered a case of champagne as incentive, and duly paid up when Reines and Cowan observed neutrino interactions in 1956.

Beta decay held a further surprise. It had long been accepted that mirror symmetry (or parity) should be an exact property of nature; this is equivalent to saying that nature does not distinguish between left and right in an absolute way. This is indeed the case for phenomena controlled by the gravitational, strong, and electromagnetic interactions, but in 1956 mirror symmetry was discovered to be broken in a maximal way by Wu and Ambler (following an earlier suggestion by Lee and Yang) in the beta decay of cobalt into nickel. The particular feature of the weak interaction which exhibits this is that only "left-handed" neutrinos and "right-handed" antineutrinos can take part. Like electrons, neutrinos have intrinsic spin, but, unlike electrons, they can only spin one way. Neutrinos spin counter-clockwise and antineutrinos spin clockwise.

The Reines and Cowan experiment was carried out at the Savannah River nuclear reactor, where more than 10^{12} antineutrinos per square centimetre emerged each second, created by the by-products of nuclear fission taking place in the reactor. It was this enormous concentration of antineutrinos that made it possible to observe their interaction, despite the fact that they interact so weakly with matter. Neutrinos are also produced in some of the decays of elementary particles such as the pion. A relatively more controlled and higher energy source of these neutrinos became available in the early 1960s with the construction of one of a new generation of accelerators, the Alternating Gradient

Synchrotron at Brookhaven. Protons could be collided with a solid target to produce an intense beam of charged pions. Some of the pions then decay, and the resulting neutrinos can then be separated out by passing the residual beam through a large quantity of absorber which filters out everything but the neutrinos. High energy neutrino beams are a standard part of today's particle physics programme.

6. Hadrons

The pion, discovered by Powell *et al.* in 1947, was the first of a multiplicity of particles, a veritable zoo, unearthed by physicists in the subsequent 20 years, initially using cosmic rays but then under the much more controlled conditions pertaining at the accelerators built during this period. (The pion itself was first produced in an accelerator experiment at Berkeley in 1948.) As the number of particles proliferated, attempts were made to organise them into families with common properties. The main divisions were into leptons, *i.e.* particles, such as electrons and neutrinos, which do not experience the strong interaction, and hadrons, *i.e.* particles, such as protons and neutrons, which do feel the strong interaction. Hadrons were subdivided into two categories by their spin: mesons (such as the pion) have integer spin 0, 1, 2,...; baryons (such as the proton) have half odd-integer spin 1/2, 3/2, 5/2,...

Among the cosmic ray debris were hadrons whose unusual properties in production and decay caused them to be dubbed "strange" particles. Among these were mesons, somewhat heavier than pions, the electrically charged "kaons", K^+ and K^-, and the two neutral varieties K^0 and $\bar{K}^0$, all with masses of about 500 MeV/c^2. Although the discovery of the K^0 is usually attributed to Rochester and Butler, who found it in cosmic rays in 1947, it subsequently turned out that they were seen, but not identified, in cosmic ray studies by Leprince-Ringuet in 1944. In any reaction initiated by pions or nucleons, the strange kaons were always produced in partnership with other strange particles such as the lambda or sigma, which are more massive than the proton and neutron and are now recognised as strange baryons.

To explain all this, Gell-Mann, Nakano and Nishijima proposed in 1953 that there exists a property of matter called "strangeness" which is conserved in strong interactions. A pion or a proton has no strangeness. If we assign strangeness +1 to the K^+, and if K^+ is produced in a strong interaction, then another particle with strangeness -1 must be produced in association with it so that total strangeness is conserved. In this way, the strangeness of all particles can be deduced from a study of which reactions are, or are not, observed. Strangeness is an example of an "intrinsic quantum number", *i.e.* one of the defining labels attached to a particle. Strangeness is conserved in electromagnetic as well as in strong interactions, but need not be conserved in weak ones.

In the early 1960s development of large bubble chambers and advances in scanning, measuring and data analysis permitted detailed studies of relatively complex events. These studies showed that there exist a large number of new "elementary" particles which decay so fast that they do not leave a visible track in the detectors. Their existence can be surmised from summing up the energies of their decay products, or can be

37

deduced from the fact that the probability of an interaction between two particles is greatly enhanced at certain energies. This great proliferation of new particles led physicists to question the accepted concept of what constitutes an elementary particle.

7.   The light quarks

As the number of particles increased, a pattern of regularity emerged, and the apparent complexity was shown to be the result of an underlying simplicity. In 1960-61 Gell-Mann and Ne'eman did for subnuclear particles what Mendeleev had done for the atomic elements in 1869. They realised that the mesons grouped naturally into octets and the baryons into octets and decuplets: the patterns of the "Eightfold Way". The decuplet had an entry missing: a baryon of strangeness −3 and negative charge, the Ω^-, with a mass predicted to be 1 680 MeV/c^2. In 1963 at the Brookhaven National Laboratory and subsequently at CERN, the predicted particle, with the requisite properties and a mass of 1 679 MeV/c^2, was found.

It was quickly realised that the patterns of the Eightfold Way implied a more fundamental layer of matter, just as the periodic table was explained by the discovery that atoms consisted of electrons circling a nucleus. In 1964 Gell-Mann and Zweig independently observed that the patterns of the Eightfold Way would arise naturally if all the particles were built from just three varieties of quarks. Two of these, the "up" and "down" quarks (u and d for short) build the hadrons with zero strangeness. Strange hadrons contain a third variety, the "strange" quark (or s). Quarks are unusual, as they have electrical charges that are fractions of the proton's charge: the u quark has charge +2/3, the d and the s quarks each have charge −1/3. Like the electron, quarks have spin 1/2. The structure of matter as known today is shown in Figure 1.

The up, down and strange properties are collectively referred to as the "flavours" of the quarks. For every flavour of quark there is a corresponding antiquark having the same mass and spin as its quark counterpart but possessing the opposite sign of strangeness and charge. Thus the strange antiquark, denoted by $\bar{s}$, has charge +1/3 and strangeness +1. Baryons are formed out of three quarks and mesons out of a quark and an antiquark. For example, a proton is formed from two u quarks and one d quark (uud), and a neutron is formed from one u quark and two d quarks (udd). This is illustrated in Figure 2. The multiplicity of observed baryons and mesons is the result of an underlying symmetry. If the symmetry were exact then all the baryons in a given octet or decuplet, for example, would have the same mass. That they do not means that the symmetry is broken, in this case by the s quark having a greater mass than the u and d quarks. Despite this, the symmetry is still apparent and provides an invaluable ordering and simplifying effect on the variety of observed particles and their interactions (see Boxes 2 and 3).

Figure 1. **The structure of matter as known today**

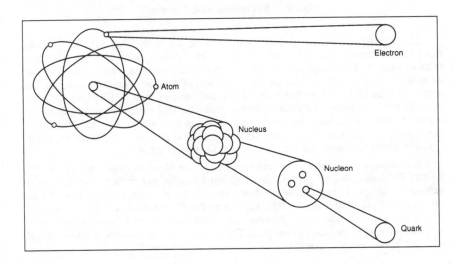

Source: Author.

Figure 2. **The quark constituents of nucleons and pions**

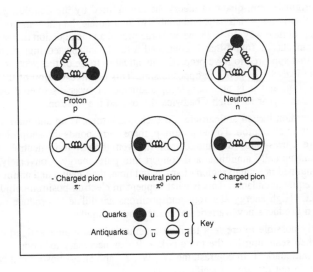

Source: Author.

Box 2. Fermions and Bosons

The matter particles, *i.e.* the six quarks and the six leptons, all have an intrinsic angular momentum, called spin. It is as if they are like spinning balls, although this is not an exact analogy. The matter particles always spin at the same rate, which is why we use the term "intrinsic". In the quantum world, the unit of angular momentum is $\hbar$, and the matter particles all have spin $\frac{1}{2}\,\hbar$ where $\hbar = h/2\pi$. Particles with half-integral spins are called fermions and obey the Pauli exclusion principle. This states that no two fermions can simultaneously occupy precisely the same quantum state (*i.e.* have identical values of momentum, charge, spin, etc., in the same region of space). It is the Pauli exclusion principle which is responsible for the chemical identity of all atoms.

The carriers of the three forces relevant for particle physics *i.e.* the photon, the W^+, W^- and Z^0 bosons, and the gluons, also have intrinsic spin, but in their case the spin is integral, *i.e.* simply $\hbar$. Particles with integral spins 0, $\hbar$, $2\hbar$, ... are called bosons and do not obey the Pauli exclusion principle. There is no limit to the number of bosons that can exist simultaneously in the same quantum state, and this allows an assembly of bosons to act coherently, as in the case of laser light. The Higgs boson, as a boson, also has integral spin, but in its case, the spin is 0.

Box 3. Antimatter

In its original form, quantum theory, as exemplified by the Schrödinger equation, was both explicitly non-relativistic and unable to incorporate the phenomenon of electron spin in an appropriate way. In 1928, when Dirac produced the equation that satisfied both these requirements, he found that it contained a third, unexpected ingredient. Particles had to exist that appeared to have properties identical to those of the electron, apart from one: the electrical charge is of the opposite sign. This positive version of the electron is called a positron, and it is an example of antimatter. It was discovered by Anderson in 1932, the same year in which Chadwick discovered the neutron.

Every fermion has an antiparticle. To each quark there is an antiquark and to each lepton there is an antilepton. To the proton, p, there corresponds the antiproton, $\bar{p}$; to the neutron there corresponds the antineutron $\bar{n}$, even though it is neutral. Matter and antimatter can mutually annihilate and convert into pure energy. Conversely, if enough energy is contained in a small region of space and time, then matter and antimatter will be produced in equal abundance. This is what happens in electron-positron colliders such as LEP and SLC. High energy electrons and positrons annihilate to produce pure energy, which in turn produces new varieties of matter and antimatter.

It is only possible to create a fermion together with a similar antiparticle. Thus at Fermilab, when searching for the top quark, t, it was necessary to produce at the same time the top antiquark, $\bar{t}$. In contrast, the W^+, W^-, Z^0, and Higgs bosons can be produced singly. They do not require a specific partner.

8. Experimental evidence for quarks

Individual quarks have never been seen, so how do we know that they are real and not simply some convenient mathematical fiction? The answer came as the result of a series of experiments at SLAC from 1968 and at CERN from 1970, which gave direct evidence for quarks trapped inside the proton and neutron. The significance of these experiments parallels Rutherford's 1911 discovery of the atomic nucleus. Since the early 1950s the proton was known to have an internal structure, and after the quark hypothesis appeared in 1964 it was natural to suggest that it is the quarks moving inside the proton that give it its size. At the energy of 20 GeV available at SLAC from 1968, electrons can resolve objects appreciably smaller than the size of the proton, and so are an ideal tool for investigating their structure.

By studying the way in which the electron is scattered by the proton it is possible to determine the distribution of charge inside the proton. If the charge is evenly spread throughout the proton, the electrons will pass through with little deflection. On the other hand, if the charge is localised on the quarks, then the electron will occasionally pass close to one of these concentrations of charge and be strongly deflected. It is a direct analogue of Rutherford's alpha-particle experiments that revealed the atomic nucleus in 1909. Just such violent collisions were seen at SLAC, and the distribution of electrons showed that the proton is indeed built from objects with spin 1/2. Comparison of these results with similar experiments at CERN, where neutrinos were used as probes in place of electrons, proved that these constituents have electrical charges which are 2/3 and −1/3 of the proton's charge. Agreement with the quark model deduced from the Eightfold Way patterns was complete. The experiments also showed the presence of electrically neutral particles, gluons, which are now known to be the carriers of the force that attracts quarks together, forming the proton. This was a very welcome discovery, as it indicated that the experiments were beginning to reveal how the proton is held together and not just what it is made of.

There were two other discoveries which were paradoxical at the time. First, the quarks appear to be very light, less than 1/3 of the mass of the proton; and second, they appear to be almost free inside the proton as if they are hardly constrained at all. This paradox was seminal for the subsequent development of the theory of quantum chromodynamics, or QCD, which has all of the desirable properties of apparently free, yet permanently bound quarks. This property goes by the names of "asymptotic freedom" and "confinement". As a crude analogy, the quarks can be considered as held together by a slightly slack elastic band. They are apparently free, but after being struck they recoil and the elastic tightens, preventing them from escaping. If the quarks are struck too hard the elastic may become so stretched that it snaps. The energy released in the break creates a new quark-antiquark pair, and so mesons are formed but not free quarks.

9. The heavy quarks

A fourth flavour of quark with a charge of +2/3, charm (or c), was proposed in 1964 by Bjorken and Glashow, and independently by Hara and Maki; the idea was developed in 1970 by Glashow, Iliopoulos and Maiani, and discovered at Brookhaven and SLAC in 1974 with the dramatic appearance of the J/ψ meson, a bound state of c and c̄. This was something entirely new: a meson substantially heavier than anything seen before (more than three times the mass of the proton) and unexpectedly stable. Within two weeks of the discovery of the J/ψ, an even more massive particle, the ψ', had been found at SLAC. Just as hydrogen exists in excited levels, so can states formed from a charmed quark and a charmed antiquark, and during 1975 a whole spectroscopy of states built in this way was discovered at SLAC. These are called states of "charmonium". To complete the picture, it was necessary to find evidence for mesons built from a charmed quark and one of the u, d or s quarks. These charmed mesons were more elusive, and it was not until the summer of 1976 that the evidence for charmed mesons became unambiguous.

In 1977 a fifth quark with a charge of −1/3, "bottom" or "beauty" (or b), appeared at Fermilab in the form of a b b̄ bound state, the Υ (upsilon). Finally, again at Fermilab, the sixth quark, "top" (or t), which has a charge of +2/3, was found in 1994. Some t t̄ pairs have been identified in very high energy p p̄ collisions by detecting the W bosons and b (or b̄) quarks emitted in their decay. Each successive quark is more massive than its predecessors: u and d have masses of 0.3 GeV/c^2, i.e. about one-third of the proton mass; s has a mass of 0.5 GeV/c^2; c has a mass of 1.5 GeV/c^2; b has a mass of 4.8 GeV/c^2; the top has a measured mass of 174 ± 20 GeV/c^2, in good agreement with the predicted value of about 160 GeV/c^2. This diversity of mass is not understood, and it is clear that the underlying symmetry, which there undoubtedly is, must be badly broken.

10. Leptons

The story of the leptons is much less dramatic than that of the quarks. The second lepton, the muon (or μ), was discovered by Anderson and Neddermeyer in 1936 (and independently by Street and Stevenson) while searching for Yukawa's pion. Although the muon has many of the characteristics of a "heavy electron" (it is about 200 times the mass of the electron) it is a distinct particle with a unique "muon" label, just as the electron has a unique "electron" label. We know that they must be different, because the energetically allowed decay of the muon to an electron plus a photon (which would be very rapid if the muon were just a heavy electron) simply does not happen. A third lepton, the tau (or τ), with a mass nearly twice that of the proton, was discovered at SLAC in 1975 and confirmed at DESY in 1976. Again, this lepton has its own unique "tau" label and is not simply a heavy version of the electron or muon.

When positively charged pions decay, they produce either an antimuon and a neutrino or a positron and a neutrino, although the decays involving the antimuon are by far the more prevalent. In 1963, using the neutrino beam at the Accelerating Gradient Synchrotron (AGS) at Brookhaven, a group led by Lederman, Schwartz, and Steinberger proved that the neutrinos produced in the two different decay modes of the charged pions are different neutrinos, ν_μ and ν_e. If there were only one type of neutrino, $i.e.$ if ν_μ and ν_e were the same, then when the neutrino beam interacts with matter, equal numbers of electrons and muons would be produced in the interaction. On the other hand, if the neutrinos were distinct, then, as the neutrino beam is composed primarily of ν_μ, the products of the interaction should contain many more muons than electrons. The antineutrinos produced in reactors from beta decays would then be antiparticles of ν_e. After some 10^{14} neutrinos had traversed their apparatus, Lederman, Steinberger and Schwartz found 51 events with muons and none with electrons, thereby conclusively proving that ν_μ and ν_e are distinct, just as the muon and electron are distinct. Each lepton has its own associated neutrino, which carries an "electron", "muon", or "tau" label as appropriate. How the neutrinos "know" their separate identities is not yet understood, but the pairing of each electron with its own neutrino appears to be a fundamental part of nature's pattern. All leptons have spin 1/2.

11. Families

The Universe in which we live today requires only the proton, the neutron, the electron and its associated neutrino, and, as the nucleons are not fundamental objects, they should be replaced by their constituents, the up and down quarks. These lepton and quark pairs are known today as the "first family" of elementary particles. Nature then repeats itself, with a "second family" consisting of the muon and the muon neutrino, the charm quark and the strange quark. There is a third repetition, the "third family" consisting of the tau and the tau neutrino, the top quark and the bottom quark. Direct evidence for the existence of top, hitherto a theoretical conjecture, was announced in May 1994 by the CDF experiment at Fermilab. Its mass is within the range predicted from LEP data using the Standard Model. This structure is illustrated in Figure 3. From experiments on the SLC at SLAC and on LEP at CERN we know that there are no more than three such families. Experimental evidence for this conclusion is presented in Figure 4, which shows LEP data on the Z^0 peak. Recognition of the family patterns for quarks and leptons is very much like the discovery of the Eightfold Way for hadrons and Mendeleev's periodic table for atoms, although so far no hint has emerged of an analogous simplifying substructure.

The individuality of the different families is exemplified by the mode of operation of the weak interaction. In the lepton sector the weak interaction operates entirely within each family, connecting e with ν_e, μ with ν_μ and τ with ν_τ. In the quark sector it operates primarily within each family, connecting u with d, c with s, and t with b. However, it does cross families, albeit at a much lower level, for example, connecting s with u and b with c and u. The decays of the τ lepton and the b quark are shown diagramatically in Figure 5. We do not know the reason for this distinction between quarks and leptons, and determining and interpreting the parameters of the "weak mixing matrix", which describes these cross-family connections qualitatively, are central to understanding it.

43

Figure 3. The family constituents of matter and the carriers of the forces between them

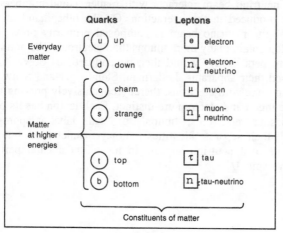
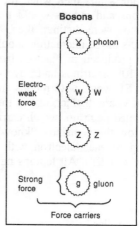

Note: The universe in which we live is composed entirely of the first family.
Source: Author.

Figure 4. Experimental evidence for three neutrinos

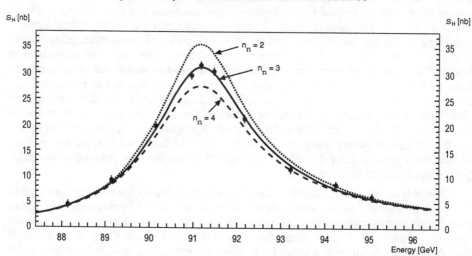

Note: The data from the LEP show the cross-section for Z^0 production in e^+e^- annihilation and the theoretical expectation for 2, 3 and 4 neutrinos.
Source: Author.

Figure 5. Decay of the tau lepton and the b quark, illustrating the difference
between the lepton and quark sectors of the weak interaction

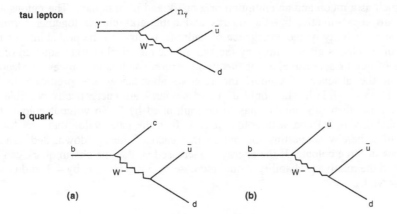

Note: For the b quark, diagram (a) is 100 times stronger than diagram (b).
Source: Author.

12. Colour

Initially, very little was known about the forces that act on quarks beyond the fact that quarks cluster either in threes (baryons) or as quark with antiquark (mesons), but in no other way. The first real clue was a paradox concerning the Pauli exclusion principle. Particles with spin 1/2 obey this principle, which prohibits having more than one particle of a given kind in the same state of energy and spin. It is this principle that controls the structure of atoms, forcing electrons into particular orbital states and thereby creating the pattern noted by Mendeleev. In nuclear physics, the principle allows at most two protons and two neutrons in the lowest energy state; this is precisely the configuration of the alpha particle and its source of stability. The same principle should apply to quarks, and this implies that the Ω^-, consisting as it does of three identical quarks in the same state, should not exist, yet this is directly contravened by observations.

Soon after the idea of quarks was proposed in 1964, Greenberg and, later, Han and Nambu, suggested that, to avoid the problem, quarks would have to possess a new property to distinguish otherwise identical quarks. This property is called colour, and it is in many ways similar to electrical charge except that it occurs in three varieties: let's say, red, yellow and blue. Instead of simply positive or negative charge, as is the case for electrical charge, there are positive and negative red, yellow or blue colours. Quarks carry positive colour charges and antiquarks have the corresponding negative colour charges. Each of the three strange quarks in the Ω has a different colour; therefore, they are not identical, and the Ω^- can exist.

45

Direct independent evidence for colour was obtained in an experiment at Frascati in 1970 and at CEA in 1971, and was confirmed, more precisely, at SLAC in 1972. When high energy electrons and positrons annihilate, their energy is converted into other forms of matter, such as a muon and an antimuon or a quark and its antiquark. The muons and antimuons are seen directly, but the quarks and antiquarks cluster together forming hadrons. The probability of the emergence of hadrons, relative to the probability of the emergence of muons, is given simply by the sum of the electrical charges squared of all the flavours of quarks accessible, as the electron-positron annihilation process is electromagnetic. In the absence of colour, the relative abundances are predicted to be $[(2/3)^2 + (-1/3)^2 + (-1/3)^2]$, when only u, d and s quarks are energetically possible. If colour does exist, then this number has to be multiplied by 3. So without colour, the relative probability is 2/3 and with colour it is 2. It is the latter value that was found experimentally, thereby supporting the notion that each of the up, down, and strange quarks comes in three colours. If the energy is increased so that the charm quark can be accessed, then the relative probability should increase by $3 \times (2/3)^2$, *i.e.* by 4/3, and this is what is observed.

13. Quantum chromodynamics

Quarks, and hadrons containing quarks, all experience the strong nuclear force, whereas electrons and neutrinos do not. As quarks possess the property of colour, but electrons and neutrinos do not, it is reasonable to suggest that colour is the source of the forces acting between quarks. In electrostatics, like charges repel and unlike charges attract. The analogy for quarks is immediate: like colours repel and opposite colours attract. This provides a natural explanation for the existence of mesons: just as positive and negative electrical charges attract to form neutral atoms, so do positive and negative colours carried by quark and antiquark attract to form uncoloured mesons. In electrostatics, two positive charges are always "like charges" and repel. For quarks, two like colours always repel, but what about two different colours, *e.g.* a red quark and a blue quark? It turns out that different colours can attract one another but they do so less intensely than the opposite colours of quark and antiquark. Thus, a red quark and a blue quark, say, can mutually attract, and the attraction is maximised if, in turn, they cluster with a yellow quark. All three separate pairs attract one another, to form the three-quark clusters of the baryons. Baryons formed in this way necessarily contain three quarks, each with a different colour; this is the picture Greenberg invented as an explanation of the Pauli principle paradox. All of the known hadrons, both mesons and baryons, are uncoloured, *i.e.* the colour of the constituent quarks or antiquarks exactly cancel each other.

In the mathematical theory of quantum chromodynamics (QCD), colour generates quark forces in much the same way as electrical charge generates electromagnetic forces in the mathematical theory of quantum electrodynamics (QED). In quantum electrodynamics, the carrier of the force is the photon. In quantum chromodynamics, the carrier of the force is the gluon. The structures of QED and QCD are very similar, but there is one essential difference. The photon is electrically neutral but the gluon carries colour charge. This means that gluons can interact with each other directly, while photons cannot. It is believed that it is this additional interaction that gives rise to the property of confinement. At distances of appreciably less than 10^{-15} m, the force between two coloured quarks is almost non-existent, but as they are separated the force increases rapidly and tends towards infinity, permanently confining the quarks in colourless clusters. Dynamic evidence for the reality of gluons and confirmation that they have spin 1, like photons, was established at DESY in 1979 (see Box 4).

Box 4. The Four Fundamental Forces

The gravitational force acts between all particles, pulling matter together, and it is of infinite range. It is the binding force of the solar system and galaxies and controls the large-scale structure of the Universe. It is exceedingly feeble in atomic and particle physics, and on the microscopic scale it only becomes important at distances of the order of 10^{-35} m or, equivalently, at energies of the order of 10^{19} GeV. This is far beyond the reach of present technology and even four orders of magnitude beyond the energies at which it is believed that the strong, electromagnetic, and weak forces unify.

The electromagnetic force acts between electrically charged particles. It is the binding force of atoms, the negatively charged electrons being held around a positively charged nucleus. It is also responsible for binding atoms into molecules, liquids, and solids. Like the gravitational force, its range is infinite, but as all matter is electrically neutral, its effective range for many purposes is on the scale of the atom. The electromagnetic force is carried by photons.

The weak force causes radioactive decay of some nuclei. A neutron breaks up, emitting an electron and an antineutrino, and becomes a proton. Seen in this way, the weak force appears appreciably weaker than the electromagnetic force. However, this results directly from the huge mass of its carriers – the W^+, W^-, and Z^0 particles – which obscures the fact that the intrinsic strength of the weak force is comparable to that of the electromagnetic force, and, at sufficiently high energy, when the mass of the carrier no longer matters, the two forces no longer appear very different from each other. This is the basis for uniting them into a single theory of the "electroweak" force, which combines electromagnetism and the weak force within it.

The original concept of the strong force was that it attracts protons and neutrons and binds them into nuclei. It is of short range, and is believed to be the remnant of a more powerful colour force acting on quarks inside the protons and neutrons. We now know that quarks exist as charged constituents of protons and neutrons, and experiments show that some electrically neutral particles (gluons) are present in addition to quarks. These gluons are the carriers of the force that attracts quarks together to form protons and neutrons. A property, termed "colour", is carried by quarks (but not by electrons): it is the strong force equivalent of the electrical charge for the electromagnetic force. As photons are to electrical charge, so are gluons to colour. At the quark level the strong force is simple. The strong nuclear force between neutrons and protons is a complicated manifestation of the more fundamental colour force acting between their quark constituents, just as intramolecular forces are complicated manifestations of the more fundamental electromagnetic force acting on the atomic constituents, the electrons.

Figure 6.   **Two-jet and three-jet production in e$^+$ e$^-$ annihilation**

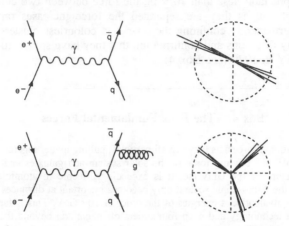

Note: The diagrams on the left illustrate the processes of quark-antiquark production with gluon radiation (three jets). The diagrams on the right show schematically what is observed experimentally: the lines represent particle tracks. Experiments like this proved the existence of the gluon.
Source: Author.

Although quarks and gluons cannot be seen directly, there is dramatic evidence of their existence in high energy interactions through the production of jets: highly collimated streams of particles, such as pions, along the line of the emerging, but unseen, quark or gluon. This is illustrated schematically in Figure 6. There have been many successful predictions of quark and gluon interactions at high energies, including the production and properties of jets and the way in which the quarks and gluons are distributed inside the proton. A notable feature of QCD is that the strength of the coupling between the quarks and gluons is a dynamic quantity. It is not a constant, but varies in a very well defined way as a function of kinematic variables, decreasing with increasing energy. This "running of the coupling constant" has been verified experimentally with remarkable accuracy.

14. Quark-gluon plasma

One particularly intriguing prediction of QCD is that by colliding energetic heavy nuclei such as lead, conditions of density and temperature can be achieved which replicate those that existed some fractions of a second after the Big Bang. In these tiny bubbles of exceedingly hot and dense matter, we expect to be able to observe, very briefly, a deconfinement transition in which quarks and gluons would no longer be confined inside "particle bags" but would move freely in larger volumes forming a quark-gluon plasma.

Significant differences between nucleon-nucleon collisions and nucleus-nucleus collisions have been observed in experiments with relativistic heavy-ion beams at

Brookhaven and CERN. In head-on nucleus-nucleus collisions, there is unambiguous evidence for the creation of a high-temperature, high-energy-density fireball, appreciably larger in physical extent than the original colliding nuclei. Associated phenomena, such as increased production of strange particles and decreased production of charmonium, are compatible with the creation of a quark-gluon plasma. However, it cannot be excluded that the observations are due to some extreme state of "conventional" nuclear matter rather than to the entirely novel form of matter of the quark-gluon plasma. To resolve this issue, nucleus-nucleus collisions at higher energies are required; they will be available in the fixed-target lead-ion beam programme at CERN, at RHIC (Brookhaven), and in the relativistic heavy-ion option at LHC (see Box 5 and Figure 7).

Box 5.   The Uncertainty Principle and Particle Interactions

It is not possible to measure both the position and the momentum of a particle at the subatomic level with arbitrary precision. We may "see" an electron by shining light on it and detecting the scattered radiation (photons), but as we do so, the electron recoils and changes its momentum because the energy of the photon required to "see" the electron is comparable to the rest-mass energy of the electron. This does not happen for macroscopic objects, whose momentum is not measurably affected.

If the position of a particle is only known to be within some distance Δr, then its momentum must be indeterminate by at least an amount Δp given by $\Delta p . \Delta r = \hbar$. Here $\hbar = h/2\pi$, where h is Planck's constant: $\hbar = 6.6 \times 10^{-22}$ MeV.s. A similar principle applies to time and energy. The uncertainty in energy at a given time implies that energy conservation can be "violated" over sufficiently short time scales. For example, an electron can radiate energy in the form of photons in apparent violation of energy conservation, provided that energy is reabsorbed by another electron within a sufficiently short time. The heavier the exchanged particle, the more the energy that must be "borrowed" to create it and hence the shorter the time and the distance over which it can travel.

It is customary to draw a Feynman diagram to represent this. Some examples are shown in Figure 7. For example, in the top diagram, an electron is represented by a straight line and a photon by a wiggly line, and time runs from left to right. The diagram represents an electron coming along and emitting one photon which is in turn absorbed by another electron, the photon transferring energy from one electron to the other in the process. The photon has effectively transmitted a force whose origin is the original electron some distance away. In this sense we say that photons mediate electromagnetic forces.

We view all forces as being due to the exchange of particles. Each of the three fundamental forces with which particle physics is concerned has its own carrier:

- strong colour force: the gluon;
- electromagnetic force: the photon;
- weak force: the W^+, W^- bosons ("charged" weak current) and the Z^0 boson ("neutral" weak current).

Figure 7. **Feynman diagrams for the electromagnetic strong and weak interactions**

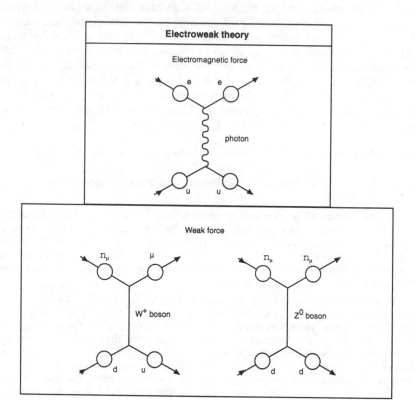

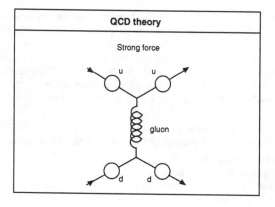

Source: Author.

15. Exotic mesons

The many successes of QCD place its validity beyond doubt; nonetheless, it does present us with a possible conundrum. We have seen how the hadrons, or "bags of quarks", can be subdivided into two types: baryons, such as the proton and neutron, containing three quarks, and mesons, such as the pi-meson, containing a quark and an antiquark. Almost all known hadrons appear to fall into one of those two categories. Why are there no baryons composed of five quarks; or mesons composed of two quarks and two antiquarks; or indeed particles composed entirely of gluons, the so-called glueballs; or even mixed hybrid mesons composed of a quark, antiquark and a gluon? Because gluons can interact with each other, bound states involving them explicitly are particularly appealing, and their existence is predicted by "lattice-gauge calculations" (the particular computational technique required for bound-state calculations in QCD). Despite extensive searches, no unambiguous evidence for such "exotic" states has been seen, although candidates have been proposed and the search continues. While reasonable explanations for the failure to observe "exotics" can be advanced, it would be encouraging to identify them.

16. CP violation

Soon after it was discovered that the weak interaction violated parity P, it was found that it did not preserve charge conjugation C in the decay of the charged K mesons either. It was possible that these two symmetry violations cancelled each other out so that the combined CP symmetry would be preserved by the weak interaction. To test CP violation it is necessary to find physical states to which one can assign a definite value of CP. These are provided by linear combinations of K^0 and $\bar{K}^0$ (which themselves do not have a well-defined value of CP: the operation of CP converts K^0 into $\bar{K}^0$ and *vice versa*). The states which are "seen" by the weak interaction are the linear combinations $K_1^0 = (K^0 + \bar{K}^0)/\sqrt{2}$ and $K_2^0 = (K^0 - \bar{K}^0)/\sqrt{2}$ which have values of CP of +1 and −1, respectively. The weak interaction acts on these states, not on the neutral K mesons produced by the strong interactions.

If CP is conserved in weak interactions, then K_1^0 can decay only to states with a positive value of CP, such as two pions, while K_2^0 can decay only to states with a negative value of CP, such as three pions. This gives rise to very different mean lifetimes for the two states: 0.9×10^{-10} s for K_1^0 and 5.2×10^{-8} s for K_2^0 which allows the latter to be separated from the former in a beam of neutral K mesons. In 1964, Christenson, Cronin, Fitch, and Turlay tested the hypothesis that weak interactions conserve CP symmetry exactly by observing a beam of K_2^0 and looking for any decay into two pions. They found that this occurred twice in every 1 000 K_2^0 decays and concluded that CP symmetry is not preserved exactly by the weak interaction.

The violation of CP symmetry has profound theoretical consequences which are still not fully understood. It might be that the weak interaction really does conserve CP and that the violation is due to some new "superweak" force. If, however, the origin of

CP violation is indeed due to the weak interaction, then evidence for this "direct" CP violation should be found in K meson decays. The experimental evidence is ambiguous: an experiment at CERN claimed to have found some indication, but an equivalent experiment at Fermilab obtained a null result. Groups at both laboratories are preparing appreciably more sensitive experiments on K meson decay. The decays of B mesons (particles composed of a b quark and a light antiquark) are expected to provide a much better "laboratory" for the study of CP violation, and this is part of the impetus for B meson factories. The presence of CP violating effects is one of the requirements for explaining our matter-dominated Universe (*i.e.* one in which there are no large concentration of antimatter) and so is of fundamental cosmological importance (see Box 6).

Box 6. Discrete Symmetries

Physicists employ the concept of symmetry extensively, and particle physicists are no exception. Symmetries provide the most fundamental explanation for the way things behave, *i.e.* for the laws of physics. The great power of symmetries is the intimate connection between them and invariance principles or conservation laws.

The three most important discrete symmetries are those of parity, P; charge conjugation, C; and time reversal, T.

If parity, or mirror symmetry, were an exact property of nature, it would be impossible to tell whether a film of an experiment has been made directly or by filming the view in a mirror. This is equivalent to saying that nature does not distinguish between left and right in an absolute way. This is indeed the case for phenomena controlled by the strong and electromagnetic interactions, but in 1956 mirror symmetry was discovered to be broken in weak interactions in the famous experiment by Wu and Ambler on the radioactive beta decay of cobalt into nickel. The particular feature of the weak interaction which exhibits this is that only "left-handed" neutrinos and "right-handed" antineutrinos can take part. Like electrons, neutrinos have intrinsic spin but unlike electrons they can only spin one way. Neutrinos spin counter-clockwise and antineutrinos spin clockwise.

Charge conjugation symmetry, C, relates to the interchange of particles with their antiparticles. This symmetry means that if physical laws predict the behaviour of a set of particles, they will predict exactly the same behaviour for the corresponding set of antiparticles. For example, the collision between an electron and a proton will look precisely the same as a collision between a positron and an antiproton. The symmetry applies also to the antiparticles of fermions with no electric charge, such as the neutron. The interaction of a proton and a neutron is the same as that of an antiproton and an antineutron.

The last of the three discrete symmetries, time reversal T, connects a process with that obtained by running backwards in time. Symmetry under "time reversal" implies that if any system can evolve from a given initial state to some final state, then it is

(continued on next page)

(continued)

possible to start from that final state and re-enter the initial state by reversing the direction of motion of all the components of the system.

There are no fundamental reasons for supposing that these individual symmetries should be preserved by the forces of nature, but there are very good reasons to think that the application of all three simultaneously, CPT symmetry, is absolutely exact. This means that for any process, its mirror image, antiparticle, time-reversed process will look exactly like the original. The consequences of this CPT theorem are that particles and antiparticles should have exactly the same masses and lifetimes, and that if an individual symmetry, *e.g.* P, or a combined pair, *e.g.* CP, are broken, there must be a compensating violation in the remaining symmetries (or symmetry) to cancel it.

Only the weak interaction is known to break any of these discrete symmetries, violating both P (in a maximal way) and the combination CP, the latter in the decay of the neutral kaons. Although it occurs at a very small level, CP violation has profound theoretical consequences. CP violation opens up the possibility of finding a mechanism for converting the early matter-antimatter Universe into a matter-dominated one and so is a crucial area of study. CP violation effects should be much more prevalent in the decays of the B meson than in the decays of the kaon, which explains the enthusiasm for B factories.

17. The Glashow-Weinberg-Salam model

The first observation of the weak interaction was in nuclear beta decay, which converts a neutron (udd) into a proton (uud) with the production of an electron and an antineutrino. At the quark level, a down quark is converted to an up quark: d becomes $u + e^- + \bar{\nu}_e$. In a theory formulated in 1933, Fermi assumed that the change in the charge of the neutron when it is transformed into a proton was caused by the emission of an electron and an antineutrino at a point. However, as early as 1938, Klein suggested that a spin 1 particle (the "W boson") should mediate this decay, and in 1957 Schwinger extended this idea and attempted to build a unified model of weak and electromagnetic forces with W^+, W^- bosons and the photon. This model had promising features but also some flaws, the most obvious one being that the W^+ and W^- both interact with neutrinos but photons do not. The apparent weakness of the weak interaction relative to the electromagnetic interaction could be explained by giving the W^+ and W^- sufficient mass, but this had to be 10 to 100 times the mass of the proton.

In 1961, Glashow suggested a solution, which was fully formulated by Weinberg in 1967 and Salam in 1968. The W^+ and W^- are the charged members of a triplet of bosons which also has a neutral member W^0: they couple to matter with a common strength g_2. There is a fourth particle, an electrically neutral B^0: this couples to matter with a different strength g_1. The W^0 and B^0 combine to form two new neutral states: one which must be more massive than the W^+ and W^-, which is the one we know as the Z^0;

and one which must be less massive than the W⁺ and W⁻ and can in fact be massless, *i.e.* it is the photon.

To make this work, Weinberg and Salam had to invoke the "Higgs mechanism", developed in 1964 by Higgs, quite independently of attempts to build theories of the weak interaction. It provided the key that Glashow did not have. It requires adding a new interaction with a massive spinless particle, the Higgs boson. Unfortunately, it is not possible to specify theoretically what this mass should be. The theoretical respectability of the whole structure was established by 't Hooft in 1971.

Prior to the development of these ideas, it had been thought that all weak interactions were of the "charged current" variety, *i.e.* the charge of a quark or lepton is changed by the weak interaction: a d quark ($-1/3$) becomes a u quark ($+2/3$), an electron (-1) becomes a neutrino (0) with the carrier of the weak force being a charged W boson. With the Z^0 now present in the theory, "neutral currents", *i.e.* weak interactions in which the charge of a quark or a lepton is unchanged, were predicted. Neutral current events were discovered in neutrino interactions at CERN in 1973 and confirmed at Fermilab in the following year. The data made it possible to predict the masses of the $W^\pm$ and the Z^0, at about 80 GeV/c^2 and 90 GeV/c^2 (*i.e.* between 90 and 100 times the mass of the proton). After converting the SPS accelerator into a proton-antiproton collider to increase the energy by a factor of 20, the $W^\pm$ and Z^0 were discovered at CERN in 1983 by a team led by Rubbia and confirmed, simultaneously, by another group led by Darriulat. Their masses are now measured to be approximately 80.2 and 91.1 GeV/c^2, respectively.

The Glashow-Weinberg-Salam model is now the undisputed theory of weak and electromagnetic interactions. Extensive tests of the model in high energy neutrino interactions, in neutrino-electron scattering at both low and high energies, in parity-violating effects in atomic transitions, in spin-dependent asymmetries in electron-deuteron scattering and now most comprehensively in complementary experiments at SLC and at LEP have all served to confirm, with ever increasing accuracy, the validity of the model in our present energy domain.

18. The Standard Model

Our current understanding is encompassed by the "Standard Model", which provides the simplest theoretical framework. It divides matter particles into two categories: quarks, some of which are the constituents of protons and neutrons, and hence nuclei; and leptons, such as electrons, which, together with nuclei create atoms, and neutrinos. There are four known interactions between these fundamental particles: the strong nuclear force which holds quarks in nucleons and nucleons in nuclei; the electromagnetic force which holds electrons in atoms and atoms in molecules; the weak force, which is responsible for much radioactivity and controls the energy processes in solar-type stars; and gravity which is ultimately responsible for all the large-scale structures we see in the Universe. So far, no confirmed results from particle physics experiments actually conflict with the Standard Model, but the model has a large number of unsatisfactory features that raise more questions than they answer. The word "model" is chosen advisedly: it is a model, not a theory, although it makes possible the performance of a large number of remarkably accurate calculations.

We have seen that the matter particles – six quarks and six leptons – come in three "families", each containing two quarks (each of which comes in three "colours") and two leptons and encompassing a natural symmetry. The first family consists of the up and down quarks, the electron and its accompanying neutrino, which, together, form the Universe we observe today. The second family contains the strange and charm quarks, the muon and its neutrino. The third family consists of the top and bottom quarks, the tau lepton and its neutrino. We know from SLC and LEP that three, and only three, such families exist. The as yet undetected ingredient of the Standard Model is the tau neutrino. However, persuasive evidence for the tau neutrino has been available for some time. The heavy top quark was discovered only recently, in 1994, at Fermilab. Because it has the required properties and a mass consistent with prediction, it confirms that the Standard Model is the correct description of nature in our present energy regime.

The Standard Model uses what are called gauge theories to describe three of the four forces that particles feel: the strong, the electromagnetic and the weak interactions. These theories are based on an underlying symmetry, or invariance principle, corresponding to the conservation of some fundamental quantity such as electrical charge, and describe the interactions in terms of gauge bosons, which are the carriers of the force. The electromagnetic force is carried by photons, the weak force by the neutral Z^0 boson and the charged W^+, W^- bosons, and the strong force by neutral gluons. The unification of the electromagnetic theory with the theory of weak interactions to form the unified electroweak theory requires the existence of the Higgs boson, which is certainly massive, with an experimental lower limit of about 60 GeV/c^2 and a theoretical upper limit of several hundred GeV/c^2. Theorists would like to unify the three gauge theories of the Standard Model into one gauge theory: a grand unified theory or GUT, for which there are several candidates. This would require unifying the electroweak theory with the theory of strong interactions, quantum chromodynamics.

19. Limitations of the Standard Model

Although the predictions of the Standard Model agree remarkably well with experiment, it cannot be regarded as the last word on elementary particles and their interactions. Many of their intrinsic properties, such as the quantum numbers that describe charge and spin, cannot be predicted and have to be put in "by hand". The Standard Model contains at least 20 such parameters. They include the masses of the quarks and leptons and of the W and Z bosons, as well as the strengths of the forces. How do quarks, leptons and gauge bosons acquire mass, and why should there be such a wide range of masses? Why does the charge on the proton, which is built from quarks, so perfectly equal the charge on the electron, which is not built from quarks? The Standard Model doesn't predict how many families of particles there are, nor why the Universe is apparently dominated by matter, rather than possessing a symmetry between matter and antimatter, as we might expect from a starting point of pure energy. Even neutrinos, which have been studied in the laboratory for more than 30 years, are enigmatic particles. Why are neutrinos left-handed and antineutrinos right-handed? Do neutrinos have mass, and if so does this explain the apparent deficit of neutrinos observed from the nuclear reactions which power the sun? If they do have mass, are they the constituents of the "hot" dark matter possibly required by cosmology?

The masses of the quarks, leptons and bosons provide us with a real puzzle. Although the Standard Model categorises them neatly and allows us to list many of their properties by simply related values, there is no apparent pattern among their masses. In one sense our situation resembles that of Mendeleev when he classified the chemical elements into the periodic table which incorporated their empirical properties, although it took quantum mechanics and Bohr's theory of atomic structure to explain why it worked. Or it is analogous to the underlying quark structure, which explained the pattern of hadrons. In another sense, we are not so fortunate, as these new elements of physics, unlike those of chemistry or of the hadrons, display no regularity in their masses. The tau lepton is about 17 times heavier than the muon and 3 491 times heavier than the electron. Equally mysterious, but different, ratios are found among the masses of the quarks. Why do quarks and the charged leptons, and the W^+, W^-, Z^0 gauge bosons have mass, while photons and gluons are massless? The masses of the neutrinos are certainly far smaller than those of the related charged leptons, but are they massless? The regularity of pattern must imply some hidden, deeper symmetry of nature. However, this new symmetry must be badly broken in a way that is not obvious to us, unlike the symmetry of the u, d, s quarks, which was broken simply because the s quark is more massive than the u and d quarks.

As a first step to understanding the masses, it is essential to have a mechanism for generating mass; within the Standard Model, the preferred mechanism is the interaction of the Higgs boson. Other possibilities, such as the possibility that the Higgs boson might not be elementary but made of other particles – for example a bound state of a top quark and a top antiquark or of a W^+ and W^- pair owing to a strong interaction between them, or the possibility that the quarks themselves may have a substructure consisting of new particles – cannot yet be ruled out, but they appear increasingly unlikely.

The Standard Model does not tell us what the mass of the Higgs boson should be, although limits on it can be inferred from experiment. Unfortunately, its mass is correlated with that of the top quark, which is not yet accurately known. Both masses enter into calculations of the relative strengths of the weak and electromagnetic forces and affect this ratio slightly. This is because both the top quark and the Higgs boson can participate in the decays of the Z^0 as "virtual" particles because of the uncertainty principle of quantum mechanics. Comparing theory and experiment puts constraints on these masses, and best estimates from the LEP data for the top quark mass are between 140 and 185 GeV/c^2 and for the Higgs boson are between 60 GeV/c^2 and 1 TeV/c^2, with a strong preference for the lower end of this range. Evidence obtained at Fermilab for the top quark gives a mass of 174 ± 20 GeV/c^2, which is compatible with the results from CERN. Within the presently accessible energy range, the key to tying down the Higgs mass more precisely, short of actually discovering it, is a more accurate measurement of the $W^±$ mass, which we can expect from Fermilab and LEP-II.

Finally, there are intrinsic theoretical difficulties if the calculations are taken to very high energies. In many ways the situation is akin to that of the Fermi model of weak interactions, which provided an accurate description of low energy phenomena but contained unacceptable features in a higher energy domain. The solution to that problem led ultimately to the unification of electromagnetic and weak interactions in the Glashow-Weinberg-Salam model, the prediction of the W^+, W^- and Z^0 bosons and their triumphant discovery precisely where they were anticipated. Is a new, underlying, simplifying and unifying interaction the solution again?

20. Tomorrow

Any consistent theoretical extension of the Standard Model is likely to involve a higher degree of symmetry than we currently have, a grand unification. One important concept is supersymmetry, which postulates a kind of mirror world to the Standard Model: every fermion of the Standard Model has a supersymmetric boson partner and every boson of the Standard Model has a supersymmetric fermion partner. This creation is not so *ad hoc* as it might seem, as the severe technical problems of the Standard Model at high energies are thereby automatically solved. The lightest neutral supersymmetric particles (which are nonetheless very heavy) are stable and are candidates for the "cold" constituent of cosmological dark matter.

Although supersymmetric particles have not been observed, can we adduce any evidence for them? A strong hint for unification comes from the relative strengths of the strong, electromagnetic and weak forces. We have noted that the strength of the quark-gluon interaction decreases as the energy increases, and the interquark forces weaken significantly at high energies. This type of variation is not unique to QCD but is a property common to all gauge theories. For example, the strength of the electromagnetic force increases with increasing energy. Conventionally, the strength of the electromagnetic interaction at low energies is measured in terms involving the charge on the electron, and which has the value 1/137. On the same scale, the strength of the strong interaction is about 1/5. Extrapolating to the huge energy of 10^{15} GeV, both coincide at about the value of 1/42. If electroweak theory and quantum chromodynamics are to be joined into a grand unified theory, then at sufficiently high energy the weak, electromagnetic, and strong forces should be of equal strength: the theory should be completely symmetric. If only the particles of the Standard Model are included in the calculations, then all three forces do not coincide at a single point. Adding supersymmetric particles to the Standard Model changes the rate at which the strengths of the forces approach each other as the energy is increased, and theory and experiment are brought into agreement within 1 per cent. This is illustrated in Figure 8 and sets the scale of supersymmetry at around 300 GeV. A bonus of grand unified theories based on supersymmetry is that they predict three families of matter particles.

The idea that symmetrical behaviour can be apparent at high energy (or temperature) but concealed at a lower energy (or temperature) is a familiar one in physics. Magnetism provides an example very close, both conceptually and mathematically, to grand unification. At high temperatures the atomic spins which are the source of magnetism are

Figure 8. **Extrapolation to high energies of the strengths of the weak, electromagnetic, and strong interactions from present data, showing how the introduction of supersymmetry (SUSY) changes the slopes and permits grand unification**

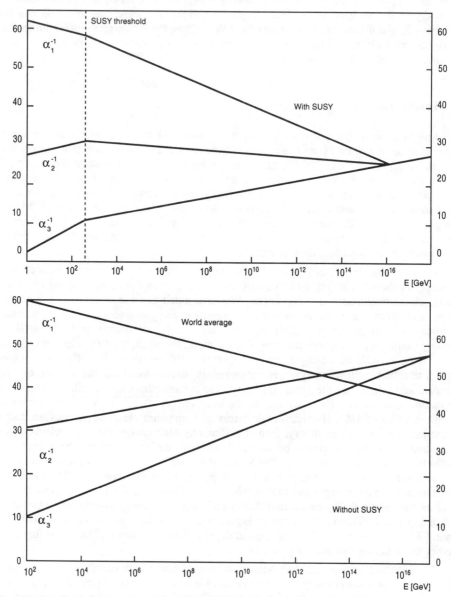

Source: Author.

randomly ordered, and the system is isotropic, *i.e.* it is fully symmetric, no one direction is preferred to another, and no magnetism ensues. As the temperature is lowered, interatomic forces assert themselves, the atoms align, and magnetism appears. There is now a preferred direction, defined by the north and south poles, and the original full symmetry has been lost even although some well-defined symmetry remains. Thus, the concept that the Universe possessed, immediately after the Big Bang (see Box 7 and Figure 9), an innate symmetry which was concealed as it cooled is fully in accord with other physical experience.

Box 7. The Big Bang, and History of the Universe

In 1964, two astronomers, Penzias and Wilson, were using a radio antenna in Holmdel, New Jersey to detect radio waves coming from outer space. To their surprise, they found a background noise similar to the static that can interfere with the reception of a normal radio broadcast. This background noise was constant, unvarying and came from all directions. This Cosmic Microwave Background, as it is now known, is a radio signal with a frequency spectrum corresponding to a temperature of only 3 K. Its source is everywhere, pervading the whole Universe. We believe that the Universe erupted in an incredibly hot fireball, the Big Bang, which is the source of the expanding Universe we observe today. Matter and electromagnetic radiation were created in the Big Bang, and as the Universe expanded, the matter clumped to form stars and galaxies and the radiation cooled. The 3 K background radiation discovered by Penzias and Wilson is the cold remnant of that once hot emission, a direct relic of the Big Bang, and the precise measurement of its temperature provides the key to establishing the temperature of the early Universe.

At 10^{-43} seconds after the Big Bang, with temperatures of 10^{32} K and energies of 10^{19} GeV, gravity was strong. Understanding these very early moments requires a theory of quantum gravity, which does not yet exist. By 10^{-37} seconds, with temperatures of about 10^{29} K and energies of about 10^{16} GeV the dominant effects came from the unified strong, electromagnetic, and weak forces. By 10^{-33} seconds, with temperatures of about 10^{27} K and energies of about 10^{14} GeV, the predominance of matter over antimatter was established, and the strong interaction was no longer united with the electroweak interaction. By 10^{-9} seconds, with temperatures of about 10^{15} K and energies of 10^{2} GeV, the electroweak interaction began to separate into the distinct electromagnetic and weak interactions. This corresponds to the exploratory limit of present accelerators. By 10^{-5} seconds, with temperatures of about 10^{13} K and energies of about 1 GeV, quarks and gluons disappeared as separate entities and condensed into protons and neutrons. At 100 seconds, with temperatures of about 10^{9} K and energies of about 100 keV, nucleosynthesis started, and deuterium and helium were created. At 10^{6} years, with temperatures of about 10^{3} K and energies of about 1/10 eV, electrons combined with nuclei to form electrically neutral atoms and photons decoupled from matter. The Cosmic Microwave Background radiation originates from this point, and optical and radio astronomy cannot see back beyond this epoch. The temporal evolution of the Universe is indicated in Figure 9.

Figure 9. Temporal evolution of the Universe from the Big Bang

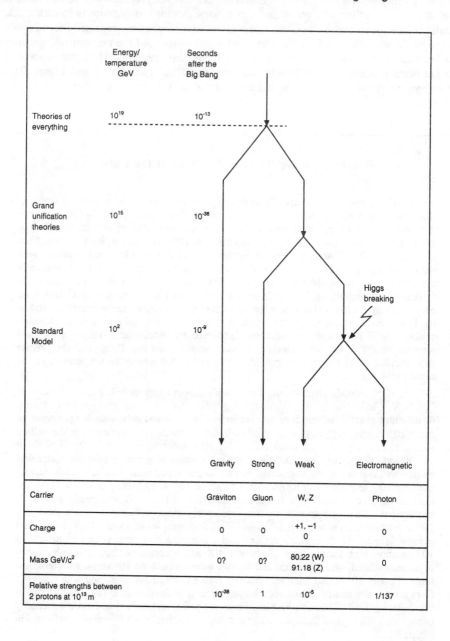

	Gravity	Strong	Weak	Electromagnetic
Carrier	Graviton	Gluon	W, Z	Photon
Charge	0	0	+1, −1 0	0
Mass GeV/c^2	0?	0?	80.22 (W) 91.18 (Z)	0
Relative strengths between 2 protons at 10^{13} m	10^{-38}	1	10^{-5}	1/137

Source: Author.

60

The weak force plays a key role in investigating the three families, for it is only through the weak force that the heavier quarks and leptons can decay to the lighter stable states of the first family. Studies of such decays have already revealed an inherent lack of symmetry in the weak force, called CP violation, which may have a bearing on the observed domination of matter over antimatter in the Universe. The three requirements for the development of an asymmetric universe, rather than one in which there is symmetry between matter and antimatter, were initially formulated by Sakharov and are: processes violating CP symmetry; processes violating the conservation of baryon number; and a state of non-thermal equilibrium. The last existed in the Big Bang; non-conservation of baryon number can be incorporated in grand unified theories and would lead to proton decay (albeit at an extremely slow rate); and CP violation does appear naturally in the Standard Model. A particularly intriguing aspect is that it require at least three families to achieve this: two would not be sufficient. The Standard Model predicts that the appropriate CP-violating effects should be visible in the decays of B mesons *i.e.* mesons which contain the bottom, or beauty, quark.

The mechanisms needed to produce the dominance of matter over antimatter require an extension of the force carriers of the Standard Model. In addition to the W^+, W^-, Z^0 and photon of the electroweak interaction and the eight gluons of the colour interaction, there are 12 new force carriers, called X and Y. Like the gluons they must carry colour charge, but unlike the gluons and the photons they must also carry electrical charges: +4/3 or –4/3 and +1/3 or –1/3. The force they transmit can change quarks into leptons, *e.g.* d (blue, –1/3) plus X (antiblue, +4/3) gives a positron, or change antiquarks into quarks, *e.g.* anti-u (–2/3) plus X (+4/3) gives u (+2/3). This force can cause the proton to decay, but as the mass of the X and Y must be that of the grand unification scale, *i.e.* about 10^{15} GeV, the probability that the proton will decay is very small. Nonetheless, the probability is finite, and experiments deep underground (put there to shield them from unwanted cosmic ray effects) are seeking, so far unsuccessfully, to observe an occasional proton decay.

21. Outstanding questions

The success of the Standard Model is astonishing, particularly when we know that, despite its elegance, it cannot be the whole story. Many of the issues raised by the Standard Model are of profound significance, as they reflect features that are essential to the observed behaviour of the Universe and are the questions to be answered by the next generation of particle accelerators:

What is the mass generation mechanism for elementary particles?

Why is the magnitude of the charge on the electron the same as that on the proton?

Why are there three, and only three, families of quarks and leptons?

Why does our present Universe require only the first family? What is the role of the other two?

Is the set of particles and forces complete?

Are quarks and leptons fundamental, or composites of a smaller number of other more fundamental entities?

Why is the weak force left-handed, *i.e.* why does the neutrino spin in only one direction?

How does the weak force break particle-antiparticle symmetry; can this explain the dominance of matter in the Universe?

Can gravity be included in a theory with the other three interactions?

Do neutrinos have mass?

What is the nature of the "dark matter" in the Universe?

Does the proton decay into lighter particles, and if not why not?

22. Summary

The development of particle physics from the late nineteenth century to the present day is outlined, highlighting some of the most significant intellectual achievements en route. As particle physicists have been able to access higher and higher energies and probe the structure of matter to smaller and smaller distances, layer after layer of complexity has been removed, revealing a unifying simplicity underneath. This has culminated in the "Standard Model" of particle physics, which successfully describes all phenomena at presently available energies in terms of a limited number of elementary particles which interact through the operation of a few fundamental forces. The creation of the Standard Model must rank as one of the great scientific achievements of this century. In studying the elementary particles and their interactions, physicists are probing backwards in time, back to the earliest moments of the Universe. Despite the astonishing success of the Standard Model, we know that it cannot be the whole story. There are many questions to be answered, ranging from why the magnitude of the charge on the proton is the same as that on the electron (when they appear to be entirely different entities) to why we live in a matter-dominated Universe (when the Big Bang would produce matter and antimatter in equal quantities). The account ends with a list of the outstanding questions, which are the ones to be answered by the next generation of particle accelerators.

Chapter 2

Use of Existing Facilities in Particle Physics

A first section presents a brief overview of most of the currently active high energy physics facilities and the main points of their physics programmes. The next section is devoted to the detectors, which sometimes represent major technical and financial contributions from the user community. These are followed by sections on the managerial and institutional modes of operation, and tables and graphs give an overall survey of the past, present, and anticipated future of each laboratory.

1. Present facilities

This overview is arranged by type of particle collision, briefly naming the installation and then describing some high points of the physics being done. The list is incomplete; it emphasises facilities that are at the energy frontier, or are "factories" of a given type of particle, or are special in some other way. The tables mentioned show which facilities are located in each of the major laboratories.

Electron-positron colliders

LEP (CERN, Geneva, Switzerland): 55 GeV electrons colliding with 55 GeV positrons, soon to be increased to 90 GeV × 90 GeV (LEP-II). The present programme is concentrating on precision studies of the Standard Model at the Z^0 peak. It takes advantage of the high luminosity of this machine, with about two million events in each of its four experiments to date. With LEP-II, it will be possible to produce $W^+ W^-$ pairs, giving a more precise determination of their mass and allowing, for the first time, a search for any anomalous interactions beyond the expectations of the Standard Model. Particle searches could be extended; in particular a Higgs boson with a mass of up to 80 GeV would be detectable.

SLC (SLAC, Stanford, CA, United States): 46 GeV electrons × 46 GeV positrons in a single pass collider. This machine serves two important purposes. It is the only prototype of a linear collider, and its lower intensity, compared to LEP, is effectively compensated, for certain purposes, by the strong longitudinal polarisation of its electron beam.

CESR (Cornell, Ithaca, NY, United States): 8 GeV electrons × 8 GeV positrons, functioning for B physics with a continual plan of accelerator improvements to increase luminosity. It lacks the kinematic advantages of the future B factories with asymmetric collisions, which allow a good event-by-event measurement of the B lifetimes, but it is producing a large amount of valuable data in the meantime.

TRISTAN (KEK, Tsukuba, Japan): 32 GeV electrons × 32 GeV positrons. The programme includes tests of the electroweak theory, investigations of quantum chromodynamics, and B physics.

BEPC (Beijing, People's Republic of China): 2.8 GeV electrons × 2.8 GeV positrons. This is the energy range for study of the c quark and the tau lepton. The high luminosity of this machine has made possible studies of greater precision.

Novosibirsk complex of e^+e^- colliders (Budker Institute of Nuclear Physics, Novosibirsk, Russian Federation): Two colliders are presently in operation:

– VEPP-4M (5.5 GeV electrons × 5.5 GeV positrons);
– VEPP-2M (0.7 GeV electrons × 0.7 GeV positrons).

Many experiments studying vector mesons (from rho meson to upsilon meson) have been carried out, including precise mass measurements using a resonant depolarisation method developed at Novosibirsk.

VEPP-2M is an interesting example of a machine designed to carry out a very specific task. The ultra-high precision measurement of the magnetic moment of the muon, the "g – 2" experiment which will be mentioned below, requires for its interpretation a precise measurement of the cross-section for the production of hadrons from these collisions over the full range, strongly weighted at the energies of this accelerator/ experiment.

Electron-proton colliders

HERA (DESY, Hamburg, Germany): 30 GeV electrons × 820 GeV protons. A unique facility allows the study of the structure of the proton through both the charged and neutral components of the electroweak current, to distances of less than 10^{-17} m, a factor of 20 less than what was possible before. This will discover or put stringent limits on quark substructure, provide sensitive tests of the Standard Model in regimes complementary to that of the electron-positron colliders, and allow searches for exotic objects.

Hadron colliders

Tevatron (Fermilab, Batavia, IL, United States): 900 GeV protons × 900 GeV antiprotons. After the discovery of the top quark, the primary goals at present are an increase in the number of such events and the precise measurement of the W boson mass, as well as detailed tests of QCD.

Fixed-target facilities

Tevatron (Fermilab, Batavia, IL, United States): 900 GeV protons. Secondary beams are used for:

 i) neutrino physics;
 ii) CP violation in K decays and rare decays;
 iii) studies of heavy quark production;
 iv) charmonium studies, using the antiproton source;
 v) hyperon decays.

Linac (SLAC, Stanford, CA, United States): 50 GeV electrons. The main focus of this programme is tests of QCD through study of scattering of polarised electrons on polarised protons and neutrons.

SPS (CERN, Geneva, Switzerland): 450 GeV protons. The programme is similar to that of the Tevatron, including the studies of neutrino mass through neutrino oscillations. This facility also supports a programme of heavy ion beams (up to lead), in a search for evidence of the creation of a quark-gluon plasma.

AGS (Brookhaven National Laboratory, NY, United States): 33 GeV protons. The main focus of the programme is the use of the facility as a factory of K mesons, in particular the study of rare decays. The experiments are now reaching the level of sensitivity where decays such as those into a pion and two neutrinos can be expected to appear. The K mesons are also used in experiments to study new configurations of quarks and gluons, such as a particle with six quarks. The muons resulting from decays in a pion beam are used to fill a special storage ring for muons, "g – 2", which will be used to measure the magnetic moment to very high precision. This facility also supports a programme of heavy ion facilities.

U-70 and UNK-600 (Protvino, Russian Federation): 70 GeV protons and 600 GeV protons. The U-70 proton synchroton is still used for meson spectroscopy and search for exotic states. Its intensity will be increased in view of its operation as injector for UNK-600, which is expected to be completed in 1998. The scientific programme of UNK-600 includes deep inelastic lepton-hadron scattering with muon and neutrino beams, the search for neutrino oscillations and for observation of the tau neutrino, and the spectroscopy of charmed particles.

PS (KEK, Tsukuba, Japan): 12 GeV protons. This machine is used for a series of experiments on K decays and interactions, to study interactions of hadrons in nuclei, and also to produce low energy muons for a variety of studies.

HERA (DESY, Hamburg, Germany): 820 GeV protons. Although not yet underway, plans have been made to use the proton ring for fixed-target studies on internal targets. 30 GeV electrons. Polarised electrons are used on a polarised gas target.

Comment on closed or redirected facilities

Many other machines have been used in the past in the research centres mentioned above. When they reached the end of their use in forefront particle physics research, a

Table 1. **Closed or redirected accelerators originally built for particle physics research**
(cutoff at 1 GeV)

Name and location of facility	Energy and type of facility	Shutdown	Re-use
Stanford	1 GeV e^-	X	
Birmingham	1 GeV p	X	
Cornell (Ithaca)	1.5 GeV e^-	X	
VEPP-II (Novosibirsk)	2 × 0.5 GeV e^+e^-	X	
Frascati	2 × 1.5 GeV e^+e^-	X	
DCI (Orsay)	2 × 1.8 GeV e^+e^-		Synchrotron radiation
COSMOTRON (Brookhaven)	3 GeV p	X	
SATURNE (Saclay)	3 GeV p		Nuclear physics
PPA (Princeton/University of Pennsylvania)	3 GeV p	X	
SPEAR (Stanford)	2 × 3.4 GeV e^+e^-		Synchrotron radiation
DESY (Hamburg)	6 GeV e^-		Injector
DORIS (DESY, Hamburg)	2 × 5 GeV e^+e^-		Synchrotron radiation
CEA (Harvard)	6 GeV e^-	X	
NINA (Daresbury)	6 GeV e^-		Synchrotron radiation
CEA Collider (Harvard)	2 × 6 GeV e^+e^-	X	
BEVATRON (Berkeley)	6 GeV p		Nuclear physics
NIMROD (Rutherford Laboratory)	6 GeV p	X	
Synchrophasotron (JINR, Dubna)	10 GeV p		Nuclear physics
ZGS (Argonne)	12 GeV p	X	
PEP (Stanford)	2 × 20 GeV e^+e^-		New storage ring (B factory)
PETRA (DESY)	2 × 24 GeV e^+e^-		Injector
PS (CERN, Geneva)	28 GeV p		Injector
ISR (CERN, Geneva)	2 × 31 GeV pp	X	
Main Ring (Fermilab)	400 GeV $p\bar{p}$		Injector/$\bar{p}$ source component
SppS Collider (CERN, Geneva)	2 × 315 GeV $p\bar{p}$	X	

Source: Author.

number of them were closed down, but many have been re-used, either as injectors for new facilities or in other fields of physics such as nuclear physics, or as sources of synchrotron radiation. Table 1 displays an incomplete list of such facilities (see also Chapter 3, Section 8).

2. Technical contributions of the users of accelerator facilities

The construction and operation of accelerators is most generally done by accelerator physicists and engineers, whose responsibility it is to provide the users with particle beams having the specific properties required for their experiments. It is then the task of these users to build, install, and operate the experimental devices specifically designed to carry out their scientific programme. Generally, several groups of physicists, from different university laboratories in various countries, collaborate on the construction of these detectors and on exploiting the data for scientific results. The corresponding costs are supported by the funding agencies of the laboratories of the user groups.

Although very large and sophisticated detectors had already been constructed in past decades for fixed-target experiments (particularly for neutrino physics during the 1970s), the advent of the last generation of high energy colliders has compelled physicists to imagine and to build huge and complicated devices in order to observe and measure the tracks of all the numerous and energetic particles produced in the beam collisions.

Whatever the nature of the colliding beams, the experiments usually employ similar methods: the detectors are designed to register the maximum of information on all the final particles emerging from the primary interactions of the beams. With the advent of modern colliders (SPEAR, DORIS, PETRA, but in particular the proton-antiproton programme at CERN in the late 1970s and early 1980s, followed by the Tevatron collider at Fermilab), large high-rate 4π detectors came into operation. Following earlier trends already partly set by the electron colliders, the detector systems have become very modular and specialised. In an assembly with cylindrical symmetry around the beam axis, the basic building blocks are essentially:

- an inner part dedicated to tracking, *i.e.* containing various devices designed to reconstruct accurately the tracks of the charged particles;
- an inner calorimeter for electrons and photons (electromagnetic calorimeter), so designated because most of these particles are absorbed therein and have their energy converted into measurable electrical signals;
- a magnet, often made with a solenoidal superconducting coil coaxial with the beam axis, which provides an intense magnetic field for precise measurement of the momentum of charged particles and of the sign of their electrical charge;
- a hadron calorimeter, generally combined with the steel yoke of the electromagnet;
- finally, muon chambers, which are track detectors located beyond the calorimeters, used to identify the eventual energetic muons produced in the beam interactions that pass through the entire equipment without being much affected.

Modularity is an important aspect, as it allows a more rational design and sharing of construction and facilitates assembly and repair. Excellent calibration has become an essential requirement. The information provided for each interaction has to be recorded in digital form (data acquisition, making use of specialised microprocessors and dedicated logical circuits), after being filtered (trigger) in order to reduce the data stream by very large factors (10^6 or more), in order to single out the rare, most important events.

The archetypes of these developments were the experiments UA1 and UA2 at CERN. Substantial improvements were provided in the conception of the four LEP experiments (ALEPH, DELPHI, L3 and OPAL) and of the two Tevatron large detectors (CDF and D Zero). Sophisticated built-in calibration systems operating in real time during data taking and new high-precision vertex detectors (silicon microstrips) are examples of these novel techniques.

The present working environment in these experiments is very favourable when compared with that of a high luminosity hadron collider. A step towards future hadron colliders has been taken by the two HERA experiments (H1 and Zeus), where bunch spacing is only 100 ns (nanoseconds), while synchrotron radiation creates a severe background problem.

At LHC, the first and foremost problem is obtaining the very high luminosity that will generate event rates of 10^9 per second or more: this requires high rejection factors at the different levels of trigger to select the rare interesting events. In addition, the inner parts of the detectors will have to support an extremely high level of radiation, requiring new radiation hard materials (hard silicium technology for electronics near the detectors). The two multi-purpose LHC detectors, ATLAS and CMS, are taking up all these challenges.

Of course, the increasing size and technical complexity of the largest detectors also imply an increase in their costs and in the number of physicists involved in their construction. The average cost – without personnel – for each LEP experiment was less than SF 100 million: it is expected to exceed SF 400 million for each LHC experiment.

3. Modes of utilisation of accelerator facilities for research

A number of different modes have been developed over many years to get the most effective research output from a given laboratory. The different sizes and the nature of the research carried out have led to a number of very different modes of utilisation. A large modern accelerator complex supporting a broad programme of research will have a number of different research modes being used at the same time. A representative sample is given below.

a) Independent users, a group or collaboration of several groups from outside the laboratory, who bring their equipment and use the beams to perform research in the accelerator laboratory. The essential resource needed from the lab is the beam itself, but, in practice, a number of other services are essential and traditionally have been provided without charge, such as rigging and internal transport, lab and office space, alignment of equipment, communication and control signals, utilities, including electricity, and the like. All these services are expensive, and form an important part of the budget of the host laboratory. Consequently, when the ICFA guidelines speak of providing a beam without charge, much more, i.e. much more cost, is usually implied. This mode is the classical type of user facility; it has its origin in an era when a much larger part of the research effort required relatively small experiments, but even at that time, other modes were in use. Laboratory policy is not uniform: for example, some laboratories allow users access to stockrooms, others do not.

b) Collaboration of independent users with a strong group inside the host laboratory (often called an in-house group). This mode also originated early in the history of the field, and addresses the problem of providing expertise and continuity. It can inhibit the creativity of users if it is too strongly encouraged, and is impractical if the number of different experiments is large.

c) Experimental facilities staffed by in-house groups, which are primarily responsible for producing the research results. This mode has been used when it was thought that the experimental facility used at an accelerator was so specialised that only a dedicated team could accomplish the research. Outside users may join the collaboration, but in a well-defined role. Specialised spectrometers at high intensity low-duty cycle accelerators are a case where this mode may be required, as in the famous and successful case of the SLAC fixed-target spectrometers.

d) Open facilities staffed by dedicated teams whose main function is to provide for smooth and reliable operation, while most of the research activity is performed by a number of outside teams. The SFM and OMEGA spectrometers at CERN were set up in this way. It is very helpful for users but rather expensive, since these facilities have to be generously staffed by professionals in order to work well.

e) Facilities open to outside users in the observer mode where the data is supplied to the users, without much intervention on their part in the detector equipment or running the experiment. The term is borrowed from astronomy/astrophysics, where data tapes are available in final form to a variety of users. Open data tapes are not common in high energy physics, because the detectors are too complex for the data to be treated by groups not intimately involved in the hardware. This policy is not due to mere proprietary attitudes on the part of the detector groups, as is indicated by the fact that this mode had a glorious history in the days when the bubble chamber played a dominant role. Bubble chamber pictures are largely self-documenting, so that little detailed information is needed beyond the pictures themselves. It is quite hard to reach the same level of transparency with a complex electronic detector.

f) Large interregional collaborations dedicated to a single full solid-angle coverage detector for which many years are required to build it, run it, and carry out the data analysis. Full coverage is a relevant detail because it requires a level of co-ordination and system integration much beyond what was needed in earlier experiments. This requires a strong organisation, but in this case it is difficult to arrange, due to the number and diversity of the members of the collaboration and their modes of support. The sheer size of these entities and their organisational complexity make them more like laboratories in their own right than ordinary experiments. The movement of high energy physics towards colliding beams and high luminosities has made this the dominant mode of execution of experiments in high energy physics. The fact that a number of these operations have been carried out successfully demonstrates that this field already has impressive skills in organising interregional activities on a large scale. We will return to the new challenges which these units present to the accelerator laboratories in which they reside.

4. Laboratory organisation and physics decision making

Here we describe the organisation of the laboratory as it concerns its utilisation for research; the overall structure, as seen by the funding and oversight bodies, is discussed elsewhere.

Even in this respect, there are major differences in the practices of different laboratories. At one extreme, the director of a single-purpose laboratory may decide on the experimental programme, with the aid of an advisory board chaired by the director. At the other extreme, there may be a programme committee with a strong chairperson from outside; the recommendations of the committee can in principle be changed by the director, but they are accepted except in rare instances, usually having to do with conflicts among different committees, or constraints on resources that the committee could not fully take into account. It has been found that, between these two extremes and over a broad range of methods, research programmes can be successfully managed; this indicates that many of the real decisions are being made at a lower level in the process. Indeed, it is at the level of review, with competition among individual proposals, that most of the decision making is done. The reviews are interactive, and suggestions by the reviewers are very often adopted by the experimenters.

Proposals for experiments arrive, largely unsolicited, from groups of independent scientists, a procedure which allows for maximum creativity. However, when a new facility is opened up, or a laboratory decides to encourage a new field of research, there will be a call for proposals. The first situation is straightforward; in the case of a call for proposals when no new element in the capability is involved, experience has shown that it is wisest to treat the responses like other new unsolicited ideas, and choose the best ones.

We have described the process as if it were always for an experiment at a pre-existing facility, but the proposal may involve a major change to an accelerator, as in the case of the proposals at CERN and Fermilab which led to the conversion of proton synchrotrons built for fixed-target experiments into proton-antiproton colliders. The next step occurs when the proposal is actually for a new accelerator.

A large experiment requires a very detailed final proposal which consumes very substantial resources in time and money, up to tens of millions of dollars. In this case, it is necessary to pass through several stages of selection, guidance, and coalescence of collaborations. These stages carry names such as: expression of interest, letter of intent, and technical design proposal. The steps occur sequentially, with six to 18 months between steps, and a very rigorous review at each step. The review may entail as much as a week-long session of five specialised panels, with four or five experts on each panel, and interim specialised review between the major steps. Generally, the collaborations themselves have internal review panels in each area to ensure that major internal decisions are made under the best possible conditions. Finally, reviews are made by bodies outside the laboratory, to ensure technical feasibility, the reliability of cost and schedule estimates, and most important, the existence of appropriate management structures. Examples exist to show that an experiment can be terminated or put on hold at any stage of this process; these requirements are not merely formal.

On the other hand, the process has shown a healthy ability to take on scientifically and technically risky projects when the balance of a programme calls for it, so that, in many cases, despite the massive amount of reviewing, the opportunity of flexibility has been maintained.

It should also be emphasised that, while it is not evident from the above description, the procedure always maintains a focus on the final physics results to be achieved. Reviewers, who are physicists rather than technicans or managers, play a leading role at every stage of the review, except perhaps the very last, which is a careful control for fiduciary responsibility.

This process, which is relatively free from the influence of political bodies or even democratic user groups, has been very successful in maintaining the scientific integrity of the programme and should not be changed.

5. Emergence of new modes of management

In the last several years, starting perhaps with the four large experiments at LEP, experiments themselves have become comparable in size and organisational complexity to whole laboratories; they thus present a new managerial challenge to the host laboratory. How can the laboratory maintain control over these units of comparable strength to itself, and what degree of control is proper? The answers are still evolving. Take a simple question as an example: Who is responsible for one of these large experiments? To put this question in sharper focus, let us take the case of CERN experiments, where, traditionally, the spokesperson does not have real executive power, but responds to decisions taken by a collaboration board formed within the experiment. In the case of the SSCL, it was decided that such an arrangement would not be acceptable to the agencies with fiduciary responsibility, who would demand accountability by the laboratory. Consequently, it was felt that the responsibility for the experiment had to be carried by some person, either the spokesperson or the project manager, who was an employee of the host laboratory. As these collaborations become still more interregional, such questions will surely be analysed further, and new structures may evolve.

6. Patterns of development and change in high energy physics accelerator laboratories

A universal and striking characteristic of these laboratories is the steady course of development of the accelerator facilities over the life of the laboratory. It is in the nature of scientific research that the experiments, and consequently the detectors, change as new knowledge is gained, but the degree of improvement and change in a given accelerator facility is perhaps greater than could have been anticipated, as Tables 2-9, which set forth the history of the facilities in a number of high energy laboratories, indicate. In many cases, the accelerators were conceived and built as rather static facilities. However, as the tables show, it has been possible not only to extend the performance characteristics by

71

Table 2. Brookhaven National Laboratory (United States)

Period	Accelerator	Parameters	Physics	Maximum number of users
1952-66	Cosmotron	3 GeV protons 10^{10} p/s	Associated production Parity violation	300
1960 to present	Alternating Gradient Synchrotron (AGS)	30 GeV protons 6×10^{13} protons/pulse 25 GeV polarised protons 2×10^{10} protons/pulse 15 GeV/ heavy ions 10^9 Au ions/pulse 10^{10} O ions/pulse	Most significant discoveries: • 2 neutrinos • CP violation • Ω^-, J, charmed baryon	800
Under construction Present-1999	Relativistic Heavy Ion Collider (RHIC)	p-Au × p-Au Au × Au 2×100 GeV/nucleon L (Au × Au) = 2×10^{26} cm^{-2} s^{-1} L (p × p) = 5×10^{32} cm^{-2} s^{-1}	Nuclear matter at extreme temperatures, quark gluon plasma, proton spin physics, other p × p, p × A physics	800

Source: Brookhaven National Laboratory (United States).

Table 3. **Cornell Laboratory of Nuclear Studies (United States)**

Period	Accelerator	Parameters	Physics	Maximum number of users
1968-71	10 GeV synchrotron	6×10^{12} e⁻/s	Electroproduction and photoproduction	50
1971-76	12 GeV synchrotron	6×10^{12} e⁻/s	Electroproduction and photoproduction	50
1979-88	CESR storage ring	8 GeV e^+e^- $L = 1 \times 10^{32}$ cm^{-2} s^{-1}	Upsilon spectroscopy B meson decays	100
1989-93	Upgraded CESR/CLEO	6 GeV e^+e^- $L = 2.9 \times 10^{32}$ cm^{-2} s^{-1}	Upsilon spectroscopy B meson decays Tau decays, charm decays	200
1994-98	High luminosity CESR/CLEO	6 GeV e^+e^- $L = 1 \times 10^{33}$ cm^{-2} s^{-1}	B meson decays	200
1999-	Continuous luminosity improvement	6 GeV e^+e^- $L = 1 \times 10^{33}$ cm^{-2} s^{-1}	B meson rare decays	200

Source: Cornell Laboratory of Nuclear Studies (United States).

Table 4. Fermi National Accelerator Laboratory (United States)

Period	Accelerator	Parameters	Physics	Maximum number of users
1972-83	Main ring	200 – 500 GeV Up to $3 \times 10^{13}/ \sim 10$ s cycle	Precision measurement of magnetic moment of hyperons High energy neutrino and muon interactions Discovery of upsilon, and hence of b quark Search for $\nu_\mu - \nu_\tau$	1 000
1984 to present	Tevatron fixed target	800 GeV Up to $2 \times 10^{13}/ \sim 60$ s cycle	Precise measurements of CP violation parameters in neutral K meson decays Measurement of the properties of charm quark particles Study of charmonium spectroscopy (charm plus anticharm)	800
1985 to present	Tevatron collider	$E_{CM} = 1.8$ TeV Up to 9×10^{30} cm^{-2} s^{-1}	Measurement of B meson production, masses and lifetimes Precision measurement of W mass and width Direct evidence for top quark	900
1994	Tevatron fixed target and collider	Continuous intensity improvements Up to $6 \times 10^{13}/ \sim 60$ s cycle (fixed target) Up to 10^{32} cm^{-2} s^{-1} (collider)	Top quark physics Study of CP violation in decays of b quarks and neutral K mesons Neutrino oscillations	1 800

Note: Fermilab has, since 1972 operated the world's highest energy hadron facility. Until 1984, the Main Ring provided proton beams up to 500 GeV, for fixed target experiments using a variety of secondary beams. Since then, the Tevatron has provided 800 GeV protons for the fixed-target programme, and 900×900 GeV p$\bar{\text{p}}$ interactions for the collider programme. Fermilab has begun construction of the Main Injector, scheduled for completion in 1998. It will replace the Main Ring as the injector into the Tevatron, significantly raising Tevatron intensities.

Source: Fermi National Laboratory (United States).

Table 5. **Stanford Linear Accelerator Center (United States)**

Period	Accelerator	Parameters	Physics	Maximum number of users
1966 to present	Linear electron accelerator	E_{beam} = 1-30 GeV I = 50 μA	Electron scattering: photoproduction Experiments with pion, kaon beams	700
1972-90	SPEAR $^+e^-$ storage ring[1]	E_{CM} = 28 GeV L = 10^{30} cm^{-2} s^{-1}	Discovery of ψ, τ and D particles, plus studies of their properties and decays	150
1980-89	PEP e^+e^- storage ring	E_{CM} = 12-29 GeV L = 6×10^{31} cm^{-2} s^{-1}	Studies of hadronic final states Measurement of B and τ properties	450
1989 to present	SLC e^+e^- linear collider	E_{CM} = 90-100 GeV L = 10^{30} cm^{-2} s^{-1}	Production and decays of Z particle, esp. with polarised electron beams	200
1991 to present	SSRL dedicated synchrotron	E_{beam} = 2.5-4 GeV L = 100 mA	Synchrotron radiation research on 10 beamlines and 27 experimental stations	650
1998-2010 (?)	PEP-II asymmetric B factory[2]	E_{CM} = 10.6 GeV L = 3×10^{33} cm^{-2} s^{-1}	Search for CP violation in B decays Studies of B, D and τ particles	500 (?)
1999-(?)	Free electron laser	E_{beam} = 20 GeV	Studies with 1Å synchrotron radiation	100 (?)

1. The SPEAR storage ring ran parasitically for synchrotron radiation research from 1974 to 1990 and was converted to a stand-alone, dedicated synchrotron radiation source in 1991 by addition of a $13 million booster ring to serve as injector.
2. The PEP ring is being converted to an asymmetric B factory during 1994-98 by addition of a second storage ring.
Source: Stanford Linear Accelerator Center (United States).

Table 6. **European Laboratory for Particle Physics (CERN)**

Period	Accelerator	Parameters	Physics	Maximum number of users
1959 to present	Proton synchrotron CPS	$E_p = 28$ GeV Up to 1.8×10^{13} p/s	Fixed target, p beams, particle spectroscopy, ν physics, antiprotons	400-500
1971-84	ISR pp collider	$E_{CM} = 62$ GeV $L = 1.4 \times 10^{32}$ cm^{-2} s^{-1}	Study of proton-proton collisions at high energy, high momentum jets, heavy quarks	330
1976 to present	SPS proton synchrotron	$E_p = 450$ GeV Up to 3.5×10^{13} p/s Heavy ions: 200 GeV/nucleon	Fixed target p beams, ν physics CP violation Ultrarelativistic heavy ions	1 700
1981 to 1990	SPS p$\bar{\text{p}}$ collider	$E_{CM} = 630$ GeV $L = 5.5 \times 10^{30}$ cm^{-2} s^{-1}	Discovery of W$^{\pm}$, Z^0, study of QCD and electroweak interactions	450
1989 to present	LEP e$^+$e$^-$ collider	$E_{CM} = 100$ GeV $L = 1.3 \times 10^{31}$ cm^{-2} s^{-1}	Study of Z^0 decays, precision tests of Standard Model	1 900
~ 1996	LEP-II e$^+$e$^-$ collider	$E_{CM} = 180$ GeV $L = 5 \times 10^{31}$ cm^{-2} s^{-1}	Search for Higgs particles Study of heavy boson couplings	1 700-2 000

Source: European Laboratory for Particle Physics (CERN).

Table 7. **Deutsches Elektronen-Synchrotron DESY (Germany)**

Period	Accelerator	Parameters	Physics	Maximum number of users
1973-93	DORIS e⁺e⁻ storage ring	$E_{CM} = 11.2$ GeV (max.) $L = 33 \times 10^{30}$ cm^{-2} s^{-1}	Study of charm and beauty quarks and tau leptons Synchrotron radiation	170
1978-86	PETRA e⁺e⁻ storage ring	$E_{CM} = 46.8$ GeV (max.) $L = 24 \times 10^{30}$ cm^{-2} s^{-1}	Discovery of gluons, test of QCD Determination of strong and electroweak couplings	400
1992 to present	HERA ep storage ring	$E_e = 30$ GeV, $E_p = 820$ GeV $E_{CM} = 314$ GeV $L = 16 \times 10^{30}$ cm^{-2} s^{-1}	Study of proton structure at large Q^2, test of QCD and electroweak interactions	900
1993 to present	DORIS III dedicated synchrotron source	$E_e = 5.3$ GeV $I = 80$ mA	Synchrotron radiation research on 21 beamlines with 39 stations	1 000

Source: DESY Laboratory (Germany).

Table 8. **KEK National Laboratory for High Energy Physics (Japan)**

Period	Accelerator	Parameters	Physics	Maximum number of users [1]
1976 to present	PS [2]	12 GeV p	Rare K decay experiment CP and T violation K decay High resolution hyper-nucleus spectroscopy	270
1982 to present	SR facility	2.5 GeV e$^+$		1 000
1986 to present	TRISTAN [3]	50-64 GeV e$^+$e$^-$	Electroweak interactions QED, QCD and jets Search for new particles Photon structure functions	300
From 1998	B factory	10.6 GeV Asymmetric e$^+$e$^-$	B physics CP violation	

1. All facilities have users from abroad.
2. The booster of the PS is also used as a neutron, muon or pion source.
3. The accumulator ring of TRISTAN is used as an SR source.
Source: KEK National Laboratory for High Energy Physics (Japan).

Table 9. **IHEP Proton Accelerator Complex (U-70 and UNK-600) (Russia)**

Period	Accelerator	Parameters	Physics	Maximum number of users
1967-72	Proton synchrotron with 100 MeV linac as injector	70 GeV protons 1×10^{12} protons/pulse Secondary π, K, μ, e, γ	Search for quarks, cross-sections	300
1972-85	Upgrade of U-70 systems	70 GeV protons 2.5×10^{12} protons/pulse Secondary π, K, μ, e, γ	Cross-sections, meson spectroscopy, rare decays	400
1986-93	Proton synchrotron with 30 MeV RFQ linac and 1.5 GeV booster	70 GeV protons 2×10^{13} protons/pulse Secondary, ν WBB	Meson spectroscopy, ν deep inelastic, polarisation	450
1994-96	Upgrading of U-70 for new physics and injection to UNK-600	70 GeV protons 5×10^{13} protons/pulse	Meson spectroscopy, search for exotics, tagged ν, and p_τ	450
1998	UNK-600 with injection from UNK-70	600 GeV protons 6×10^{14} protons/pulse 5×10^{12} protons/s	Meson spectroscopy, charm study, search for ν_τ	700
	UNK-3000	3 000 GeV protons 6×10^{14} protons/pulse		

Source: Protvino Laboratory (Russia).

large factors, but even to develop whole new applications. The particle type has been changed, as when heavy ions have been accelerated in machines built for protons. Accelerators built to produce beams for fixed-target experiments have been converted into powerful facilities for colliding beam research. This is a striking achievement when one remembers that when colliding beams were first considered, it was thought that very specialised machines would be required.

Existing accelerators have been used as injectors for new machines, with most impressive savings in cost, as compared to a new facility. The most straightforward case is that of a new ring of higher energy fed by an existing accelerator complex, but there are many others, as when a new injector provides a different type of particle to existing accelerators, or when secondary particles such as antiprotons are introduced into storage rings, to be used directly for experiments or reintroduced back into the accelerator complex that made them. Sometimes the accelerators are modified so that they have multiple concurrent uses. Thus, the Tevatron is used both as a storage ring and as a fixed-target accelerator. The SLAC linac still continues to be used for fixed-target experiments, while also having been used as an injector for SPEAR and PEP and more recently having been modified into an e^+e^- collider, the SLC.

The flexibility and adaptability of high energy accelerator complexes seem to be greater than that of any other large scientific instruments. It is easy to see that this gives a tremendous advantage to existing laboratories *vis-à-vis* new institutions. The development of human resources at the laboratory, which goes along with the continuous process of change and improvement in the facilities, increases this advantage. The complex of skills developed in a laboratory becomes difficult to assemble in a new institution.

The drive to stretch the capabilities of an accelerator complex without starting afresh has led to a mastery of the art of the accelerator that would probably not have been achieved if accelerators were developed only for single practical applications. This is a vital contribution from this type of research that is hard to quantify (see Chapter 3).

It is essential to point out here that the user community often changes, even quite radically, as an accelerator complex adds new capabilities and perhaps drops older lines of research. Sometimes a new community joins the old one for new uses of the same overall facility, as in the case of the heavy ion users at BNL and CERN, while the old community sometimes moves on to other machines. Of course, the source of funding may change at the same time. The dynamics are somewhat different for a multi-purpose laboratory like Brookhaven, as compared with the so-called single-purpose high energy physics laboratories. It is worth noting that single-purpose laboratories have almost all found it useful and feasible to accommodate research or other uses of their facilities outside their nominal field, as in the case of research with synchrotron light in almost all electron accelerator facilities, the use of muons for condensed matter research at the proton accelerators, and medical treatment and isotope production at several proton facilities.

Given the scientific power, unmatched human resources, and devoted and diverse user communities associated with these laboratories, it is no wonder that they have been very influential in the formation of policy regarding new facilities. It is often remarked that it is very difficult to launch a new international effort, owing to opposition from existing laboratories, which resist the inevitable diversion of resources which a new effort will entail. The above considerations indicate that this is not only a matter of self-interest; it also reflects very real efficiencies in the operation and extension of the well-developed laboratories. These should be taken into account in the formation of future plans, perhaps to the extent that international efforts will have to consider very carefully the possibility of a base in an existing laboratory. This is all the more true when one considers the consequences of a drastic curtailment of the activities in a successful laboratory. Abandoning a massive investment and dissipating the complex human skills developed over time is not done lightly, even setting completely aside political factors. If the facility has gone on to a new life with a new user community outside high energy physics, this problem does not arise in so serious a form.

7. Impact of change on the relations with the scientific user community

In many cases, the changes in the high energy accelerator complexes have been driven by the changing scientific needs and interests of the on-going user community. Science is always changing, as old problems are solved and new ones arise. In such cases, the users and laboratory management engage in a continuing dialogue in order to advance the research potential of a facility.

In other cases, the accelerator scientists see how they can develop a new idea efficiently or make new use of their accelerators, and have to press the users a bit to appreciate the new capability. In still other cases, a new set of scientists interested in quite different problems comes to the laboratory with a proposal showing how their science could be effectively pursued with an adaptation of the accelerator complex. The user community is very fluid, but it develops loyalties which go beyond mere regional considerations. As a practical matter, experimenters learn to use a particular facility more effectively with time. If the scientific programme requires these users to shift their research to another facility, perhaps at a distant place, attention needs to be paid to facilitating their integration, otherwise their support for the scientific benefits of a new plan will not be as warm as it could be.

8. Competition between advancing the frontier and full utilisation of existing facilities

Research programmes involving existing facilities almost always encounter this tension. The annual operating costs of large facilities are comparable to the cost per year during construction, usually within a factor of two or three. The amazing fertility of imagination and technical skill of the accelerator scientists make it possible to increase the number of different experiments that can be performed at one time as well as the number of different types of beams and experiments that are possible. The incremental costs will be relatively low, because the basic infrastructure of the accelerator complex is providing multiple functions, but the total cost of all operations surely increases.

At the same time, the history of high energy physics has usually shown that the most important advances come from new facilities, particularly those that extend the energy frontier in a given type of collision. The construction of such a facility is likely to be the most expensive part of the overall programme, and the resulting additional demand on finite resources is sure to cut into the operations of existing facilities enough so that some things which could be done are not done for financial reasons. The conflict can simply be insufficient money to run the accelerator complex as many hours as possible, or the inability to meet other demands of the on-going programme, such as small upgrades to the accelerators and detectors. The argument can then be made, and always *is* made, that the programme is not being run efficiently, since expensive facilities are lying idle during a portion of the time. Of course, this is true of most of the productive tools of society. It is the overall efficiency of the programme that matters, and this may be optimised without necessarily running the facility all the possible hours. It does seem, though, that there needs to be achieved some reasonable minimum of running time, probably at least one-quarter of the maximum possible. If this cannot be sustained, serious questions arise as to whether too many facilities are being maintained or built.

It should be noted that there is another source of tension between operating a facility and constructing a new one. Often the new facility conflicts directly with the operation of an existing one, as when a new machine goes into the same building or tunnel. There will be a substantial period of time with no beams, and this clearly has an impact on the users and on the normal life of the laboratory.

9. Analysis of the activities of existing facilities

Tables and, where available, budgetary graphs give a historical view, as seen by the laboratory itself, of the activities and resources of several laboratories (Tables 2-9, Figures 10-17). These include future projects, which are described in detail in Chapter 5.

An examination of the tables suggests several points:

- The present set of research facilities displays a high degree of variety and complementarity. Most accelerators are pursuing unique programmes and there is not a great deal of duplication.
- Many laboratories have ambitious plans for the future, though not always in high energy particle physics.
- The ability of the laboratories to re-use their older accelerators in new applications is truly impressive, showing that the investment in these machines has been more efficient than originally appeared.
- The tables illustrate the history of the application, to other fields, of accelerators built for high energy physics. It is important to note that these applications have been made possible not just by the existence of these accelerators, but also by the development of beam manipulation techniques and accelerator parameters, such as beam brightness, that go far beyond the levels originally anticipated. This is an illustration of the productive effects of the great pressure exerted on accelerator parameters by the demands of particle physics research.

Figure 10. **CERN budget (1953-97) and total expenditure (1977-94)**
(million SF)

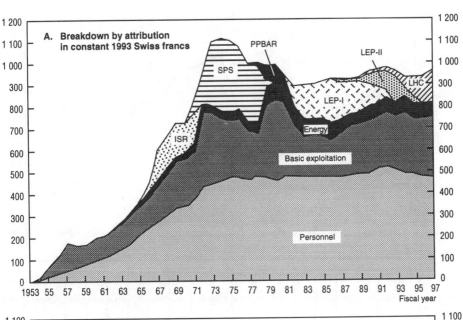

A. **Breakdown by attribution in constant 1993 Swiss francs**

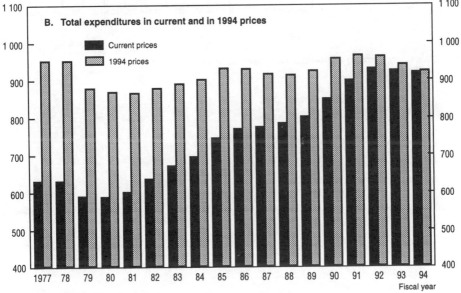

B. **Total expenditures in current and in 1994 prices**

Source: European Laboratory for Particle Physics (CERN).

Figure 11. **Budget for DESY, 1977-95**
(million DM)

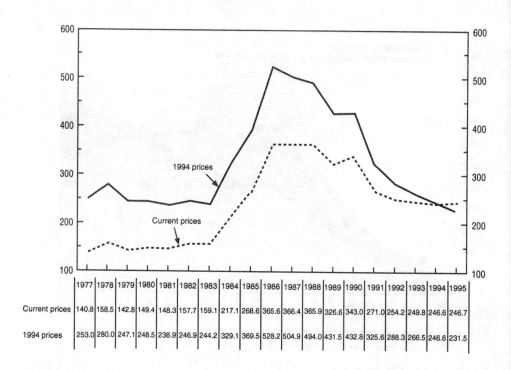

	1977	1978	1979	1980	1981	1982	1983	1984	1985	1986	1987	1988	1989	1990	1991	1992	1993	1994	1995
Current prices	140.8	158.5	142.8	149.4	148.3	157.7	159.1	217.1	268.6	365.6	366.4	365.9	326.6	343.0	271.0	254.2	249.8	246.6	246.7
1994 prices	253.0	280.0	247.1	248.5	238.9	246.9	244.2	329.1	369.5	528.2	504.9	494.0	431.5	432.8	325.6	288.3	266.5	246.6	231.5

Note: Covers only Hamburg and without synchrotron radiation.
Source: DESY (Germany).

Figure 12. **Budget for KEK, 1971-93**
(10^8 yen)

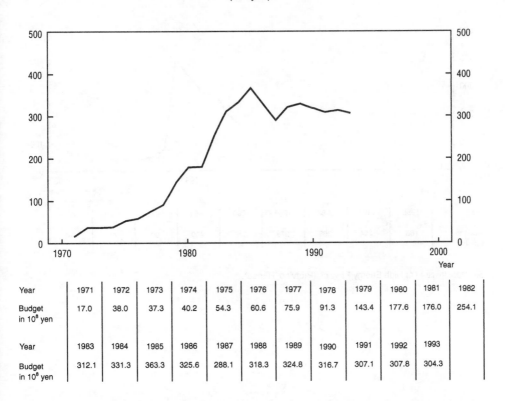

Year	1971	1972	1973	1974	1975	1976	1977	1978	1979	1980	1981	1982
Budget in 10^8 yen	17.0	38.0	37.3	40.2	54.3	60.6	75.9	91.3	143.4	177.6	176.0	254.1

Year	1983	1984	1985	1986	1987	1988	1989	1990	1991	1992	1993
Budget in 10^8 yen	312.1	331.3	363.3	325.6	288.1	318.3	324.8	316.7	307.1	307.8	304.3

Note: The budget, in current yen, includes salaries, which amount to at most 15 per cent in 1993. It includes the construction costs for the 12 GeV PS, the 2.5 GeV SR facility, and TRISTAN, as well as their operating costs. ¥ 10^8 is roughly $1 million in 1994.
Source: National Laboratory for High Energy Physics (KEK), (Japan).

Figure 13. Budget of IHEP (Serpukhov/Protvino, Russia), 1986-94
(operations and investments, million dollars)

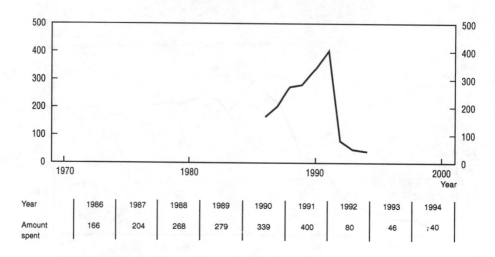

Year	1986	1987	1988	1989	1990	1991	1992	1993	1994
Amount spent	166	204	268	279	339	400	80	46	; 40

Source: Institute of High Energy Physics (Protvino, Russia).

Figure 14.   **United States high energy physics, total funding 1960-95**
(hundred million constant 1995 dollars)

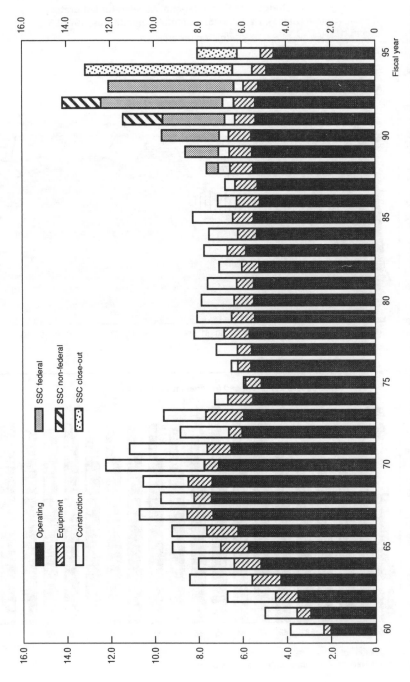

Legend:
- Operating
- Equipment
- Construction
- SSC federal
- SSC non-federal
- SSC close-out

Source: Reproduced from the High Energy Physics Advisory Panel's Report (Drell Report, May 1994), by permission of the US Department of Energy, Office of Energy Research, May 1994.

Figure 15. **Budgets of Fermi National Accelerator Laboratory (United States), FY 1968-95; and total expenditures, 1977-95**

(million dollars)

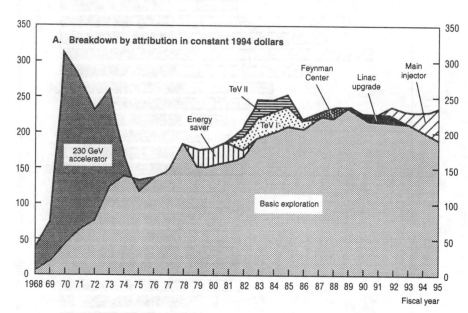

A. Breakdown by attribution in constant 1994 dollars

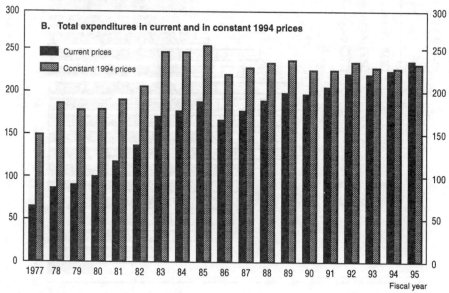

B. Total expenditures in current and in constant 1994 prices

Source: Fermi National Accelerator Laboratory (United States).

Figure 16. Budget of Stanford Linear Accelerator Center (United States), 1976-94
(million dollars)

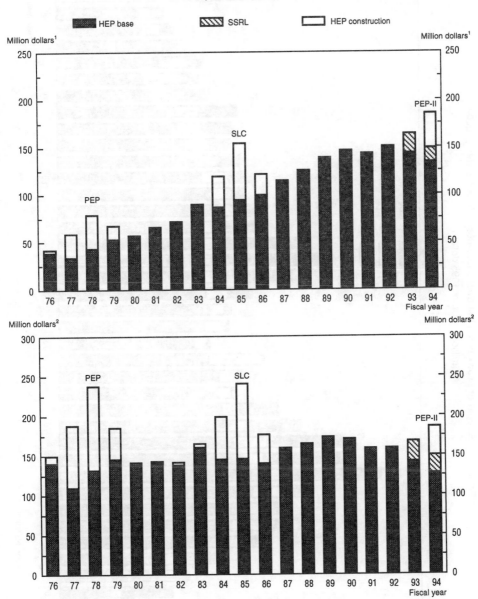

1. In current dollars.
2. In constant 1994 dollars.
Source: Standford Linear Accelerator Center (United States).

Figure 17. **Budget of Brookhaven National Laboratory (United States), 1960-95**

(million constant 1995 dollars)

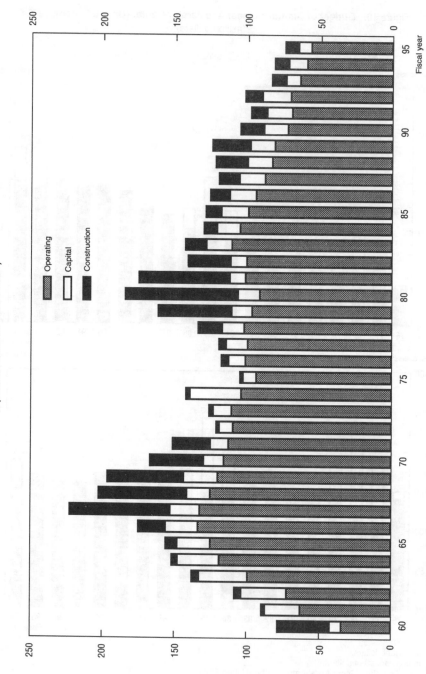

Source: Brookhaven National Laboratory (United States).

- The tables show that the substantial infrastructure in the existing laboratories and the creative imagination of their staff and users can generate a very rich programme for the future, which will require making choices. The participation of the international research community should allow these choices to be made in an optimal manner.

These points serve to demonstrate the relevance and timeliness of the present study.

10. Summary

The currently active facilities, working either at the high energy frontier or as "factories", *i.e.* with the highest possible beam intensity, are located in a relatively small number of laboratories (less than ten) with large accelerator complexes which generally support broad research programmes. Depending on the type of particle collisions, they can be classified in four categories: electron-positron colliders, electron-proton colliders, proton-antiproton colliders, and fixed-target proton accelerators. Facilities in different categories generally carry out different scientific programmes, but they are frequently able to address the same scientific question in complementary ways.

In order to carry out the most effective research, co-operation by a number of user groups has been the rule for many decades; due to the complexity and the cost of the experimental installations, the work of constructing detectors and analysing the data is usually shared by the user teams. To build and operate the huge detectors now installed at the most attractive facilities, giant collaborations have been formed, in which several hundreds of physicists join efforts. The corresponding costs, which will reach the level of several hundred million dollars for LHC experiments, are mainly supported by the funding agencies of the laboratories involved in the collaborations.

Specific management and review processes have been set up to guarantee that the focus on the final physics aims is firmly maintained, while keeping technical flexibility and financial control.

Past experience shows that the scientific community has developed an impressive capacity to make the most efficient use of former investments, by reconverting facilities no longer at the forefront of research to new applications: the 28 GeV CERN Proton Synchrotron (PS) or the Brookhaven 30 GeV Alternating Gradient Synchrotron (AGS) are striking examples.

In order to obtain the best scientific payoff from the available resources, older facilities still producing valuable results have nevertheless had to reduce their activity, or be closed, in order to make possible the construction of new instruments for a more attractive scientific programme. In the history of particle physics, difficult decisions of this sort have had to be taken.

A careful examination of the scientific programme of the present largest facilities shows their high degree of complementarity: most are unique in their category or in their energy range. Duplication in the experimental set-ups is generally limited to the minimum needed to confirm the exactness of the results.

Chapter 3

Socio-economic Benefits of Particle Physics

1. Introduction

The field of particle physics attempts to unravel the fundamental laws that have governed our Universe since its origin some 15 billion years ago. It tries to understand the fundamental constituents of all matter and the forces that govern their behaviour. While the fundamental reason for studying particle physics is to understand better the workings of the Universe, the study of particle physics also results in a number of benefits to society at large which have great potential for improving the quality of life of the world's population. Sometimes these benefits stem directly from knowledge gained about the laws of nature; more frequently they are an indirect result of the process of this research. This chapter describes some of the benefits that society derives from its support of particle physics.

Several words of caution are in order. No rigorous study quantifying the value of benefits derived from each unit of particle physics research has been made. It is not even clear that there are well-developed economic methods for doing so. To quote from one study of economic payoffs from basic research: "The eventual economic gains that will be realised on the basis of further scientific research in this field cannot meaningfully be reduced to cost-benefit estimates and thereby be compared with returns on other social investments." Thus, the arguments advanced here will be somewhat qualitative.

No claim is made that particle physics is exclusively responsible for any of the socio-economic benefits discussed below. The examples of spinoffs described in this chapter result from the use of facilities for particle physics and nuclear physics as well as synchrotron radiation sources, and often from a combination of these. In addition, the relationship between research, development and manufacturing is complex. Many different lines of research, frequently completely unrelated, often contribute to a final product or process. Progress in particle physics, as in most other physical and life sciences, is based on a fruitful interplay between ideas, experiments, and technology, and most of its successes are due to a constant and symbiotic relationship among these three areas. An exciting new result frequently gives an impetus to new ideas and initiates new technology (or improves the old). The resulting new ideas and new technology will, in turn, give rise to new or more sophisticated experiments.

The following discussion will be far from complete. Its goal is to illustrate the broad spectrum of activities that have been influenced by particle physics. The various sections attempt to categorise the benefits into several areas; the boundaries between them are somewhat arbitrary, but they are useful. Parts of Sections 6-10 are based on documentation prepared as part of the study conducted by the International Committee on Future Accelerators (ICFA) on spinoffs from particle physics and on the proceedings of a workshop on accelerator applications held at SLAC in December, 1993.

2. Particle physics and cultural values

How much should society value knowledge for the sake of knowledge alone? The answer is quite subjective and may depend on a given society's wealth and well-being. The issue is strongly related to questions such as the fraction of wealth which should be allocated to aesthetic (as opposed to purely functional) values in architecture, to conservation of natural resources, and to the preservation of the environment. The common thread here could be called social responsibility toward future generations.

Looking back three or four thousand years, we see that the intellectual achievements of past civilisations provide one of their most important legacies. The Pythagorean theorem, the Copernican-Galilean system, the Darwinian concept of evolution, the unravelling of the genetic code will all live forever. By probing the inner secrets of nature, particle physics, and basic scientific research in general, maintain this spirit of inquiry and dedication to knowledge. Although the story of the Standard Model is still incomplete, it may well be considered one of the outstanding intellectual triumphs of the twentieth century. In any case, the desire to understand the structure of matter at the most fundamental level embodies an intellectual driving force that has been central to the development of modern science and the modern mind and has been an element of cohesion in world civilisation.

Many of the results obtained during this century as part of the quest to understand nature have had profound philosophical implications. Special and general relativity, quantum mechanics, the uncertainty principle, and the wave-particle duality of matter have altered the way we think. Although all of these ideas come from the first half of the twentieth century and from investigations in areas that were the intellectual roots of particle physics, it is quite likely that current particle physics discoveries will also alter our general thought processes. The impact of particle physics on cosmology has already led to a better understanding of the origin of the Universe.

To conclude this section, we wish to make some arguments for science as a whole. The overall process of scientific enquiry fosters values that are of general benefit to society. The pursuit of science teaches the value of a critical and questioning spirit, an attitude that leads to progress not only in science but also in many other areas. It shows that most conclusions are tentative and subject to later, more precise investigation; one learns to admit that one has been wrong and that yesterday's truth no longer holds. In a time when dogmatic beliefs have led to wars and genocide, it is worthwhile promoting such attitudes.

3. Particle physics and other sciences

Particle physics brings important intellectual benefits through its impact on other sciences, but the benefits are more symbiotic than unidirectional. Probably the aspect most discussed today is the interplay between particle physics and astronomy/cosmology. The Standard Model, based on laboratory experiments, allows us to predict the behaviour of the Universe in the initial stages of its first second of existence; astronomical and astrophysical observations, such as nuclear abundance or the existence of dark matter, provide both input and new challenges to particle physics theories. The puzzle of CP (charge parity) violation, discovered and studied by particle physicists, holds the key to understanding why our Universe evolved with more matter than antimatter. The shortage of solar neutrinos may teach us about neutrino masses and mixing between lepton families.

Examples of interdependence with other sciences abound. They emphasise the unity of science and interconnections between different scientific disciplines. Gauge theories find important applications in both condensed matter physics and particle physics. The theory of quantum chromodynamics (QCD) and the predicted quark-gluon plasma are relevant to both particle and nuclear physics. The efforts to develop string theory as a theory relevant to particle physics push mathematics towards new advances. The same parameter of the electroweak theory can be determined through experiments in both atomic and particle physics, even though the energy scales probed differ by more than ten orders of magnitude.

Particle physics has had a major impact on other scientific disciplines through the technology developed in the pursuit of its intellectual goals. Much of this technology is now a standard part of the arsenal of tools used by other scientists. This will be discussed in more detail in subsequent sections.

4. Social benefits of particle physics

The focus here is on benefits outside the intellectual area. Probably one of the most important ones stems from the international nature of particle physics. Particle physics is big science and requires large and expensive facilities for most of its research. As a result, countries often have to pool their resources to build the necessary facilities, which are frequently unique in the world. Thus it has become the practice in the field that laboratories welcome as users qualified scientists from other countries. Citizens of different countries, often from different continents, frequently work together to seek an answer to a question important to all of them.

The creation of the next generation of accelerators will require international collaboration on a much greater scale. New levels of scientific and political commitment will be necessary and new problems will have to be overcome. There is awareness in the scientifically advanced nations that such a collaboration offers many challenges but also many opportunities; a number of activities are already on the way towards making such a collaboration possible.

International co-operation in particle physics has had many positive spinoffs. The European Laboratory for Particle Physics (CERN) was an early and extremely successful example of international co-operation. Founded in the 1950s, it demonstrated that people from countries that were bitter enemies less that a decade earlier could work effectively together and thus further international understanding. The success of CERN led subsequently to the formation of other European organisations, such as the European Southern Observatory (ESO), the European Molecular Biology Organisation (EMBO), and the European Synchrotron Radiation Facility (ESRF), among others. Today, established scientists, as well as students, from different continents work together at CERN, Fermilab, KEK and other large particle physics laboratories. Technical and scientific know-how is more important than the colour of one's passport. Thus, people from different cultures and different nations have an opportunity to get to know each other as human beings, frequently at a relatively early age.

In the early days of the Cold War, international particle physics conferences provided one of the first forums for contacts between Western and Eastern scientists. Many of the same scientists were subsequently involved in negotiations to limit nuclear arms and eliminate atmospheric testing. Mutual respect and familiarity deriving from those early contacts undoubtedly helped in those negotiations.

Particle physics can have other social implications. The construction and operation of laboratories centred around large accelerators are big projects, requiring many people working and interacting together. Thus, the scientists directing these projects must be attentive to social problems and their potential impact on scientific productivity. Hence a big laboratory, like other large social organisations, can be an agent for social change.

Fermilab is a case in point. It was built in the late 1960s and early 1970s in the western suburbs of Chicago, where there was strong racial discrimination. The laboratory made a special effort to recruit and train African-Americans, and thus created conditions that improved their image in the neighbouring communities. In addition, senior laboratory officials were instrumental in bringing diverse groups together and helping to open lines of communication. Gradually, barriers were overcome and equal housing and employment opportunities followed.

Finally, particle physics can contribute to a country's scientific prestige through international recognition of its successes. Of the 33 Nobel prize winners in physics in the last 15 years, 16 were particle physicists.

5. Particle physics and education

When a country pursues excellence in particle physics research, there may be many educational benefits. Early in life, children are attracted to science because it attempts to find answers to such fundamental questions as: What is the world made of? How did the Universe begin? Will it ever end? Recognising that answers to those and similar questions can be found through science, many are motivated to pursue studies in the physical sciences and mathematics. The appeal of sophisticated instrumentation, such as the giant accelerators used to study these fundamental questions, is another important factor. For many young students, particle physics represents what science is about; particle theory exemplifies the intellectual challenges of science. If seeking the answers to those questions is one of a nation's goals and thus socially rewarding, this provides an additional stimulus to scientifically oriented education. Not many of these children will end up studying particle physics, but they are more likely to obtain a solid grounding in sciences and mathematics that will leave them better prepared to cope with the modern technological world.

Large scientific installations in particle physics, such as BNL, CERN, DESY, Fermilab, KEK and SLAC can be and have been influential in fostering science at the high school or college level. Most of these laboratories now have programmes aimed at bringing high school and college students into a research environment, putting them in contact with frontier researchers, and exposing them to real research environments. At some of these laboratories, these programmes are geared towards groups, such as women or minorities, which have traditionally been underrepresented in science. In addition, many laboratories have special programmes for high school teachers, which allow them to update their own training and then transfer that knowledge to the students.

Graduate students play a vital role in particle physics. CERN has periodically plotted the age distribution of new users and has always found a pronounced peak in the 25-30 year-old group, the typical age of graduate students and postdoctoral fellows, who receive many educational benefits. They receive "hands-on" experience in the international high-tech environment of the large centres and learn from direct contact with a staff experienced in many different technologies. They also need to follow tight schedules and strict quality requirements, while facing stiff competition from competing experiments.

Most of these students and postdocs do not remain in particle physics, and most leave the academic world and research laboratories. They enter such diverse fields as the chemical or pharmaceutical industries, communications, computing and networking, the medical industry, investment banking, physics instrumentation, fusion and other energy research, the electronic components industry, and sales in many areas. This continuing outflow of students into industry constitutes the best and most efficient form of technology transfer.

Special mention should be made of accelerator physicists. These scientists, trained within the framework of particle physics activities, acquire very valuable skills that are becoming increasingly sought by other disciplines. They deserve credit for adapting particle accelerators for many applications in a variety of fields.

Many former particle physicists have started their own successful companies. Others have moved to other fields, where they have made significant contributions. Several particle physicists have received Nobel prizes for work performed in other areas, a reflection of the general value of their training.

Finally, particle physics institutes play an important role in the further training of technical personnel. The high technology used in particle physics in areas as diverse as electronics, mechanical construction, or surveying requires the acquisition of skills in the use, and sometimes the development, of the most modern instrumentation in a wide range of disciplines. When those scientists, engineers, or technicians move to private industry or government positions, they carry with them skills that they can apply in other areas.

6. Particle physics and medicine

Medicine may be the discipline that has been affected the most by particle physics research. The results of some of that research are part of the foundation of radiation medicine and very much shape practices in that field today. The primary tools of particle physics, accelerators and detectors, are also key tools in radiation and nuclear medicine; they play an important role in both diagnostics and therapy. While the tools used may not be the same as those particle physicists use today, their original concept was motivated by research in that field. Computer codes developed for particle physics find medical applications. These points, and related ones, are elaborated on below.

Particle physics results form the basis of much of what is done in radiation medicine today. For example, the Bragg peak in the energy loss curve of the stopping proton forms the basis of proton radiation therapy. The large energy release generated when π^- stops and is captured has provided the stimulus to investigate pion beams as sources of therapeutic radiation. The knowledge of the behaviour of electrons and γ-rays as they pass through matter is important for optimising radiation therapy for tumours so as to minimise exposure of healthy tissues. The same knowledge is also essential for designing shielding associated with that therapy.

Accelerators are used in medicine in two broad ways: diagnostics and therapy. In the area of diagnostics, radioisotopes play a very important role. By injecting them into living organisms and then detecting the decay (or annihilation) products, one can follow in a non-invasive way the subsequent distribution of tagged atoms and thus study various biological processes. Accelerator-produced positron-emitting radioisotopes are rapidly becoming an important tool in nuclear medicine. They have two main advantages: first, because they are short-lived they do not result in high radiation doses; and second, the annihilation γ-rays from positrons are easily detected and localised. Those favourable characteristics have given rise to the growing field of positron emission tomography (PET).

Another important accelerator-based diagnostic technique is based on the X-ray flux from synchrotron radiation. Non-invasive angiography relies on the fact that the sensitivity of contrast agents, such as iodine, can be enhanced by two or more orders of magnitude by taking the difference between a pair of exposures with two monochromatic X-ray beams, one just above and one just below the absorption edge of the agent.

Radiotherapy is an extremely important tool in the treatment of cancer. It has been estimated that 20 per cent of the adult population in the United States will receive radiation therapy. The technique aims to maximise destruction of malignant cells with minimum damage to healthy tissues. Beams of various composition and various energies are used, depending on the kind of tumour and on its location. The most common treatments still rely on X-rays or γ-rays, generally produced by linear electron accelerators, up to 40 MeV in energy. More localised energy deposition can, however, be obtained by using heavy charged particles, such as pions, protons, or heavier ions. In this way, high doses can be delivered to the malignant cells without damaging other tissue. Pion therapy is still in the experimental stage, but dedicated therapeutical proton and heavy ion facilities have recently been built. The National Cancer Advisory Board has estimated that the United States alone will eventually need 500 such facilities. Another important technique is fast neutron therapy, used to treat tumours that are resistant to X-ray treatment. Neutrons are generally produced by charged beams from cyclotrons.

Recently there has been renewed interest in boron neutron capture therapy (BNCT), a technique used to treat inoperable brain tumours, in which epithermal neutrons interact preferentially with boron that has been incorporated in a drug that selectively binds to cancerous cells. Even though BNCT now relies on neutrons from reactors, there is a general agreement that high-current accelerators are the optimal source for clinical purposes.

The last 30 years have seen a rapid development of radiation sterilisation, using electron accelerators in the MeV range. Because it can be done at room temperature, it allows sterilisation of pharmaceuticals and medical supplies that cannot tolerate heat. It can also be performed after the products have been packaged.

Use of radiation for therapeutic or diagnostic purposes almost always requires particle detection. Positron emission tomography (PET) generally uses crystal detectors to measure the two 511 keV γ-rays resulting from positron annihilation. Use of radioactive isotopes or tracers calls for detection of their decay products. Positioning and collimation of beams used for therapeutic purposes are achieved with charged particle or γ-ray detectors.

The wire chambers developed in the last 20 years for particle physics research are finding very wide use in medicine today. They offer significant advantages over photography in terms of the rate of data acquisition and the signal-to-noise ratio. In applications requiring imaging techniques, they can frequently give information equal to that obtained through photography with a radiation exposure of up to a factor of 10 less. The data acquisition techniques developed in particle physics play a very important role here.

Computer codes developed for particle research sometimes find use in medicine. Probably the best example is the electron-gamma shower (EGS) code developed through international collaboration at SLAC. This code allows detailed studies of the development of electromagnetic showers induced by electrons or γ-rays and thus is essential for designing appropriate photon radiation therapy. It is also important in designing the associated collimation and shielding.

Research in particle physics can also affect medicine indirectly. Thus, for example, research in the area of high-quality superconducting cable motivated by particle physics resulted in an increased magnetic field strength and reduced the cost per unit wire. These advances lowered the costs of magnetic resonance imaging (MRI) scanning instruments and improved the resolution through higher field strength.

Finally, one should not neglect the impact made on medicine by people trained in particle physics. Several important ideas for instrumentation and diagnostic procedures have come from particle physicists. Probably the best known example is the CAT (computerised axial tomography) scan technique, invented by Allan Cormack, a particle physicist who received the 1979 Nobel Prize for medicine for his invention. At least some of the ideas for this 3-D reconstruction of a biological organ from several X-ray scans came from challenges faced in trying to reconstruct the particle physics events recorded in stereo format on photographic film.

7. Benefits from results of particle physics research

Particle physics is a relatively esoteric science; it probes the energy domain (GeV-TeV) that is far removed from what we encounter in everyday life (eV-keV-MeV). Because of this, the results of particle physics research often take a long time to make a significant impact on our day-to-day activities, if they ever do. However, it should be remembered that the MeV energy scale was also once far removed from everyday experience. Intellectual benefits from cross-fertilisation between particle physics and other scientific disciplines have already been discussed, as has the effect of the knowledge of the behaviour of protons, neutrons, pions, electrons, and γ-rays on the practice of radiation medicine. Here we shall mention some examples that affect more closely our everyday lives.

The discovery of parity violation, inspired by unexplained puzzles in particle physics phenomena and first observed in nuclear physics and particle physics experiments, is a fundamental result and one of the foundations of the Standard Model. This violation results in an asymmetry in muon decay, the direction of emission of the electron being correlated with the spin of the muon. This fact is of considerable importance in materials science research through a technique known as "muon spin rotation". Muons are captured by the nuclei in tight orbits to give mu-mesic atoms. The study of decay muon asymmetry makes it possible to learn about the structure of materials.

The scattering of protons, deuterons, tritons and alpha particles in the 1-10 MeV range was at one time the frontier of particle physics; today this would be classified as low energy or nuclear physics. But knowledge of these interactions is crucial to the planning of fusion reactors and may in the future be the basis for generating energy.

Modern particle physics is a new science, and it may be too early to expect the results from its research to influence our lives. But intriguing suggestions have already been made. In principle at least, it is possible to use the relatively high transmission of neutrinos to "X-ray the Earth" and search for deposits of oil. If "strange matter" exists, as a number of theorists have recently speculated, it may provide a new, very compact, energy source. Whether such ideas are science fiction, pipedreams, or possibilities of future reality, only time will tell.

8. New uses for old accelerators

In order to perform new important particle physics experiments, new instrumentation frequently has to be developed. These instruments frequently advance the technological frontier in many diverse areas. The most important ones, and the ones most specific to particle physics, can be divided into two categories: accelerators and detectors. This section describes some of the uses of accelerators in fields other than particle physics, aside from medicine, discussed above.

The last six decades have witnessed an exponential growth in the energies available through the use of accelerators. Since the start of this century, the accelerator concept has evolved from tabletop instruments into giant colliders, like the 27 km LEP collider at CERN. In the process, accelerators at a given energy have become cheaper, more compact, and more reliable. The awareness of new possibilities made available by these devices has spread to other fields, both in frontier research and in applied technology. As a consequence, many accelerators have been adapted to serve other areas. The devices used today in medicine, industrial applications, element analysis or basic research in other fields may bear little resemblance to the instruments used today (or yesterday) for particle physics research. But they are the direct outgrowth of that research (or of research in the two intellectual forerunners of particle physics: nuclear and atomic physics). If it were not for the desire to understand the structure of matter at a fundamental level, it is unlikely that these tools would exist today.

It is too early to predict what role accelerators currently used in particle physics research will play in other areas in the future. We can only extrapolate from past experience. Thirty years ago, few could have predicted the broad uses of radiation from electron synchrotrons. The future is unpredictable and human ingenuity knows no bounds; we cannot state with certainty what impact current machines will have on human life in the year 2020. As one possibility, the use in the next century of the SLAC 2-mile accelerator as the key element of a free electron laser generating coherent X-rays in the 1 Å range is already being discussed.

Some of the more important current applications of accelerators are enumerated below.

Research applications

One of the most significant new research tools of the last two decades has been the electron (or positron) storage ring used as a source of synchrotron radiation. New research fields have been opened up; the availability of synchrotron radiation, with its brightness and tunability, has offered others new opportunities. The latter property finds many uses in condensed matter physics and materials science by making it possible to study atomic arrangements in many condensed matter systems.

Many areas in biology have been revolutionised by synchrotron radiation. It is now possible to study the crystallography of proteins, solve structures of large systems such as viruses, and follow the structural changes of a molecule binding to an enzyme. It is possible to study the dynamics of biological processes with time frames lasting only 10 ms. These studies provide invaluable information to pharmaceutical companies for drug design. In chemistry, synchrotron light sources make it possible to obtain unique information on oxidation of molecules, chemical bonding in solids, gases and absorbed layers, structure of complex molecules and their dynamics, and kinetics of chemical reactions.

101

Besides electron synchrotrons, other types of accelerators also play important roles in research. Proton accelerators in the range of 0.5 to 1.0 GeV are used as neutron spallation sources. Because of their neutral charge and ability to penetrate, neutrons make excellent probes for the study of condensed matter, through study of neutron scattering and neutron diffraction. These accelerators also serve as sources of pions and muons, which provide important probes for materials science through techniques such as muon spin rotation and π^+/μ^+ lattice steering.

Ion beams are used in many processes to determine the elemental composition of samples. For example, charged particle activation allows one to identify trace elements at the 1 ppb level. It is also a sensitive and fast technique for studies of wear, such as corrosion or erosion. Small spot-size ion beams can be used as an imaging device to obtain a map of the elements in the studied sample. Ion accelerators are also important in atomic physics where they are used to perform a variety of studies on complex multiparticle systems. In astrophysics and cosmology, accelerators play an important role in measuring the cross-sections of the nuclear reactions that play a key role in nuclear synthesis in stars. A recent development in this area is the use of radioactive particle beams, which are necessary to investigate some reactions.

Analysis of elements

Some of the accelerator techniques used for composition analysis and trace element detection are now used in applied sciences, archaeology, art, and even air travel security. In oil exploration, for example, neutrons, produced by the bombardment of a tritium target by deuterons, are used to activate the surrounding rocks. The γ-ray spectrum of the activated nuclei makes it possible to determine the rock composition along the well.

For archaeological dating with C^{14}, accelerator mass spectroscopy has replaced beta decay counting as the preferred method because its greater sensitivity makes it possible to perform the test with only a minute sample. Nuclear reaction activation analysis is used for studies of precious art objects. This technique has made it possible to determine the composition of ancient jewels or of pigment layers in old paintings and thus to detect forgeries or additions to works of art.

The possibility of detecting explosives is a recent development. Because explosives are nitrogen-rich, they can be detected through nitrogen activation, and compact accelerators have been designed for this application for use in airports. The explosives are detected owing to the preferential absorption of 9.172 MeV γ-rays by N^{14}; these are generated by a 1.75 MeV proton beam impinging on a C^{13} foil target.

Industrial applications

Today, a major industrial use of particle accelerators is ion implantation for manufacturing semiconductors. This technique has replaced the ion diffusion formerly used because it offers great accuracy and reproducibility for the manufacture of compact microcircuits. With this technique, ions of different types can be introduced at a desired depth, depending on the requirements of a specific circuit. The types of accelerators frequently used for this purpose are electrostatic belt generators, radio-frequency (RF) linacs, or radio-frequency quadrupoles (RFQs).

Ion implantation can also be used to modify other properties of materials besides electrical conductivity, such as surface hardness, corrosion resistance, friction coefficient, fatigue behaviour, adhesive properties, or catalytic behaviour. In the automotive industry, this technique is used to improve the performance of highly stressed components.

Radiation processing of materials is based on the fact that passing charged beams through matter can ionise and break up molecules. The resulting fragments tend to have high chemical activity, a property useful for a number of industrial processes, which tend to rely on the increased rate of polymerisation and the cross-linking of polymers. The applications are quite diverse: the manufacture of products with better mechanical, heat, or biological resistance, the creation of new materials, and the sterilisation of various products.

Radiation treatment finds many uses in the food industry. Radiation treatment of crops in storage is a more benign method of dealing with insects than toxic chemicals. Releasing males sterilised by radiation over a large area is an efficient method for dealing with pests in a natural environment. Radiation is also used to prevent the unwanted germination of crops, to suppress the activity of micro-organisms in readily perishable foods, and to sterilise processed food products to allow much longer storage times.

Synchrotron radiation, previously discussed under research applications, may soon become an important tool in the manufacture of integrated circuits. As one moves to ever more compact chips, finer and finer resolution is required. The presently used optical light technology probably cannot be pushed much beyond 0.5 μm line widths; shorter wavelengths will be required. Synchrotron radiation can offer X-ray beams suitable for lithography, by virtue of its brightness, good collimation, and high power. About ten synchrotron radiation machines suitable for this purpose, in the 0.5-1.0 GeV energy range, are presently under construction.

Micromachining is another, potentially very important, application of synchrotron radiation. Micromachines are three-dimensional microstructures such as sensors, actuators, miniature gears, and microsurgical tools ranging in size from 10 to 500 μm. Micromachinery is made with X-ray lithography, which makes it possible to make resist patterns up to 500 μm in depth. The essential synchrotron radiation properties here are the high degree of collimation and high flux at short wavelength (2-3 A).

Power engineering

The power industry offers a number of possibilities for using accelerators. The ideas are still at a rather early stage of research, but they may provide practical payoffs in the future. Some interesting concepts are: the use of heavy ion beams to achieve inertial confinement fusion; generation of ignition temperature in a magnetically confined plasma by an originally negative ion beam which is neutralised before crossing the magnetic field; conversion of ^{232}Th or ^{238}U into a fissile material by spallation neutrons; incineration of transuranic wastes from nuclear reactors by transmuting them into stable or short-life isotopes with a beam from a proton accelerator in the 1 GeV range.

Environmental protection and restoration

Accelerators have many areas of application for protecting and restoring the environment. Electron beam treatment of stack gas is being used to remove SOx and NOx from flue gas in coal-fired power plants. Only 2 per cent of the generated power is needed for this purpose. This technique can also be used for purifying the emissions from the incineration of municipal solid wastes and for cleaning up the waste water from industrial and agricultural uses. The Budker Institute of Nuclear Physics in Novosibirsk, Russia, has built about 120 electron accelerators for such environmental applications.

Pulsed electron beams are effective in destroying the toxicity of chemical wastes. The technique works by creating free radicals which in turn break chemical bonds. The products formed are considerably less toxic than the original compounds. Unlike many more conventional technologies, these techniques do not generate significant increases in the volume of waste.

9. Particle physics detectors in other fields

Modern detectors are complex pieces of apparatus, frequently several stories high, consisting of a number of different subsystems, and relying on some of the most sophisticated technology in many different fields. Several of these technologies have been invented or developed specifically to solve problems in particle physics. The subsystems that make up these detectors frequently find applications outside the field of particle physics. Some of the more important ones are described below.

Plastic or crystal particle detectors

Particle physics experiments require plastic or crystal detectors capable of producing fast signals. They also have a varying sensitivity to charged particles, neutrons, and electromagnetic radiation. They can measure, via total absorption, the energy of the incoming particles (X-rays or γ-rays, for example).

All the industrial uses of radiation that require detecting it rely on such detectors. Non-destructive testing in materials science is one use of these devices. Charged particle scattering is used to obtain information about material composition at or near the surface. X-ray fluorescence analysis gives information about atomic composition.

These particle detectors also find uses in research in other scientific fields. Medicine has already been mentioned. Neutron detectors are very important as diagnostic tools in fusion research. X-ray and γ-ray detectors play an important role in various astrophysics experiments as well as in biological research.

Tracking chambers

The need for precision and high data rates in particle physics experiments has stimulated extensive development of electronic tracking chambers over the last two decades. Resolutions of better than 100 μm are now possible and two-dimensional information on single photons can be obtained.

These developments offer many new possibilities for applications in industry and other research areas. Large-area detectors have been constructed for such applications as airport screening and inspection of large devices. Chambers have been built to study the structure of proteins by the imaging of 10 keV X-rays diffracted by their crystals. Positron cameras using the wire-chamber technology were constructed to detect 0.511 MeV annihilation γ-rays, and have made a significant impact on solid-state physics.

Electronic circuits

In parallel with the improvements in particle detectors, there have also been many advances in associated electronics. Thus, fast and precise analogue-to-digital circuits (ADCs) have been developed to measure energy and ionisation in particles. They find new applications in other fields when, for example, energies of photons have to be measured as part of trace analysis or radiography.

10. Particle physics and technology

This section discusses uses in other fields of some of the technologies developed in particle physics research. The discussion focuses on some of the most important applications, organised by the different technologies involved.

Superconducting technology

Even though superconductivity was discovered early in this century, it did not enter the realm of practical applications until some 30 years ago, shortly after type II superconductors were discovered. Superconducting technology was first applied to large magnets when the giant bubble chambers were constructed in the late 1960s. Bubble chamber magnet operation was in a DC mode, and its cable requirements were rather modest. However, when particle physicists set out to build the first accelerator magnets in the early 1970s, new cables had to be developed for the much more demanding cycling operation and for achieving magnetic fields of very high precision; the first successful cable for this purpose was developed at Rutherford Laboratory and became the basis for all subsequent accelerator magnet applications. More recently, further improvements in superconducting technology have been achieved in the Tevatron, HERA, SSC, and LHC projects. Magnetic fields have been pushed to ever-increasing values and related problems have been overcome as they have arisen. Simultaneously, the demand for unprecedented amounts of superconducting wire has stimulated the creation of new industrial capability and resulted in significantly lower costs.

These advances are now benefiting many other fields. For example, the magnets providing confinement in fusion reactors almost certainly have to be superconducting; otherwise a net energy gain would not be possible. Electric generators and electric power transmission are two areas ripe for applications of superconducting technology. While the economic climate and the oversupply of fossil fuels make this uneconomical at this time, the situation is bound to change in the future. The viability of superconducting transmission technology was already tested at Brookhaven National Laboratory in the 1970s. Large-scale storage of electricity is another area for potentially fruitful use of superconducting technology. Such systems would eliminate the need for variation in the power output of power plants and thus considerably increase their efficiency. The higher fields that are now possible with the improved cable will improve the results in nuclear magnetic resonance (NMR) spectroscopy and MRI diagnosis by providing increased analytical sensitivity.

Vacuum and surface technologies

The recent popularity of storage rings and colliders in particle physics has significantly increased the need for high-quality vacuum in accelerator applications. Pressures in the $10^{-10} - 10^{-11}$ Torr range are now routinely required. This has required developing a variety of new components, such as sputter ion pumps, all-metal valves, seals, and gauges for ultra-high vacuum (UHV) applications. In addition, it has stimulated the development of new technology for cleaning and preparing surfaces so as to render them suitable for UHV applications.

These developments have had important applications in two general areas. First, the improvements in vacuum technology themselves have had many applications in space science, fusion research, and facilities for semiconductor manufacturing. Second, the cleanliness techniques developed for UHV applications are becoming useful in other areas, such as surgery, manufacture of pharmaceuticals, and food processing. They are also making possible the development of an increasing number of new industrial manufacturing processes that take advantage of newly possible cleanliness and surface conditions.

Electrical systems

The operation of particle accelerators places demands on electrical technology that frequently exceed the state of the art. Thus, linear accelerators required the development of new klystrons, with high power and high efficiency. Some of these developments then had a beneficial impact on other areas, such as radar applications and fusion research.

Synchrotrons require very stable magnetic fields, with fluctuations not exceeding the 10^{-5} level and maintained from cycle to cycle. The powering of large accelerators from the electricity grid presents a number of challenges. A scheme of reactive power compensation had to be developed to minimise voltage fluctuations in the main grid; power converters had to be developed with high efficiency and precision of 10^{-5} or better. In addition, many of the accelerator elements need to be triggered with nanosecond accuracy. This requirement led to the development of a variety of pulse generators and great progress in industrial thyratron switches and spark gaps. Many of these developments have other promising applications in the field of high power devices.

Computing

Particle physics has pioneered the use of computers for extensive off-line analysis, simulation studies, and on-line instrument control. Work on the last of these was stimulated by the high premium placed on useful accelerator beam time and the complexity of the instrumentation both in accelerators and detectors. Many of the hardware and software systems used today in other fields had their origin in particle physics applications.

The first coincidence circuit was developed in the late 1930s to study cosmic rays and their interactions. That basic circuit is the foundation of the modern computer industry. More recently, the concepts of emulators and processor farms based on many microprocessors have been developed for uses in particle physics. Very fast, special-purpose computers have been developed by particle physicists for specific problems, such as lattice-gauge theory calculations. All of these concepts are finding applications in a variety of different industries.

Both particle physics accelerators and detectors contain a large number of complex, and sometimes geographically separated, subsystems. Transferring efficient and reliable information to and from these subsystems requires sophisticated, real-time, distributed computing systems. Those systems involve not only new hardware but also advanced, specially developed software. The applications developed for particle physics find uses in distributed databases, distributed mail systems, and the power industry. In the area of interactive hardware, great attention has been devoted in particle physics to user-friendly man-machine interfaces, to the ability to process a great deal of data, to the identification and diagnosis of possible malfunctions of different pieces of apparatus, and to the calibration of different subsystems. Clearly, these are functions that are useful far beyond particle physics.

Computer simulation has become an essential part of experimentation in particle physics. Relevant computer codes cover topics such as the performance of complex detectors, the behaviour of stored particles in storage rings, and the interaction of particles in matter. These simulation techniques have been emulated by other research fields and by industry. Some particle physics codes have direct applications in other fields, such as the EGS code in medical applications already mentioned. Hadronic shower simulation codes are important in spallation source designs and in inertial fusion research. Both of these codes are used in the design of radiation shielding.

To deal with the great deal of data arising in the course of a typical particle physics experiment, sophisticated programmes for computer analysis had to be developed. Several of the techniques involved, such as pattern recognition algorithm, statistical analysis, and chi square or likelihood fits are now used in fields such as engineering, astronomy, biology, medical research, etc. For example, the techniques developed in particle physics for extracting a rare signal out of a great deal of data have been influential in the recent work in astronomy that identified the massive, non-radiating, microlensing objects in the halo of our Galaxy.

Because particle physics collaborations often include institutions from all over the world, efficient collaboration needs to optimise exchange of data, analysis codes, and results. Network developments such as the World Wide Web developed at CERN have aroused interest among other specialised institutions.

Electronics

Particle physics has always advanced the state of the art in electronics, especially in the areas of speed and rate. This need has stimulated development (by research groups themselves and by industry) of many special circuits and systems as well as instrumentation which subsequently found applications in other areas. Some of the more important examples are data acquisition systems, logic circuits, wave form digitizers, and digital oscilloscopes.

Particle physics was one of the first fields to realise the importance of standardisation in certain areas. Crate and bus systems, such as computer-automated measurement and control (CAMAC) and the VERSA Module Eurocard (VME) originate mainly from particle physics. Today, they are widely used in all research areas and in industry.

Other technologies

Particle physics research has made an important impact on a variety of other technologies. Many techniques in mechanical engineering were either introduced or improved by the work in this field. Examples of this work include the use of unconventional materials in the construction of accelerator components and detectors, chemical polishing and hydroforming, thin-film technology, and electron-beam welding used to manufacture superconducting niobium RF cavities.

Civil engineering has profited from a number of survey instruments developed in the process of constructing large accelerators, whose components sometimes have to be aligned to a precision of 0.1 mm over a distance of kilometres. Tunnelling technology has profited from such devices as "geological radar" which makes it possible to use electromagnetic sensors to learn about the nature of the rock ahead of the boring. The technique was developed for particle accelerator tunnels.

The operation of accelerators requires a variety of instruments to measure such beam parameters as position, intensity, losses, etc. Very sophisticated and diverse instrumentation has been developed for this purpose in particle physics, and it is now used in other applications of particle beams.

11. Impact on industry

This chapter concludes with a short discussion of the general impact of doing business with particle physics research institutions on industrial firms. As noted at the outset, it is difficult, if not impossible, to measure these effects rigorously. The following comments should therefore be viewed as somewhat anecdotal.

Both CERN and KEK have attempted to evaluate the economic impact of their orders on their suppliers of high-technology products. These studies relied on questionnaires sent to vendors and thus the responses were rather subjective. However, it is clear that impact has been significant; orders have increased and costs have been reduced, owing to benefits from the required investment in new facilities, favourable publicity, subsequent growth of the market, and use of the research performed at these particle physics institutes.

The construction of the electron-positron collider (BEPC) near Beijing, China, placed very severe technological demands on the supplier industries. Chinese industries were forced to develop and produce goods which met existing Western high-technology standards. This has put them in a better position to qualify for other difficult high-technology orders.

Major suppliers of high-technology goods to particle physics research institutions tend to maintain close contact with the staff at those institutions. These take the form of frequent informal contacts, permanent presence at the institution, or participation in a formal technology transfer programme. Major electronics and computer suppliers, for instance, keep permanent personnel on site at particle physics institutes. Frequent collaborative efforts to develop new products and subsequent tests and improvements benefit both the scientific users and the industrial suppliers. Building on the expertise acquired at the particle physics laboratories, industry built the superconducting magnets for the HERA collider, and industry was to produce almost wholly the SSC magnets. In preparation, industry people spent considerable time at Brookhaven and Fermilab working with laboratory staff to build the pre-production prototypes.

A significant economic benefit undoubtedly accrues to the industrial companies through interactions that result in better informed personnel, new ideas for relevant products, and rapid feedback. To our knowledge, quantitative assessment of these benefits has not yet been undertaken.

12. Summary

The main reasons for doing particle physics research are cultural; the resultant benefits are somewhat intangible and difficult to quantify. Certainly no rigorous cost/benefit analysis is possible. Research into the fundamental workings of the Universe, such as that performed in particle physics, enriches our culture and helps to stimulate other scientific fields. Perhaps its biggest immediate impact is in the area of education: the questions the field attempts to answer, the sophisticated instrumentation it uses, and the exciting results it provides, all act as important stimuli in attracting young people into technically oriented education. A better-educated society is the end result.

But these primary reasons for doing particle physics should not distract us from appreciating the wide range of secondary benefits derived from the use, in other areas, of tools and techniques developed for the study of the fundamental nature of matter. Other scientific disciplines, medicine, computing, and the electronics industry are all much richer today because of the past investment in particle research. On the basis of the past, there is every reason to believe that this trend will continue, even though it is not possible to foresee an exact scenario.

Chapter 4

Existing Mechanisms for International Co-operation

1. Introduction

High energy physics is a basic science with a long tradition of international co-operation. Because its results are published in scientific journals, and are thus available to researchers throughout the world, collaboration on an international scale has evolved quite naturally.

As researchers have investigated smaller and smaller elements of matter, they have required larger and more sophisticated facilities. This is a function of a law of the subatomic universe, that the shorter the distance, the greater the energy, *i.e.* the larger the facility, that is needed. Thus, as accelerators and detectors grew in scale, it became increasingly necessary to draw on resources from several countries. Lately, international co-operation is so taken for granted that it is hard to find a programme that is confined to one country.

While there is a tendency to see international co-operation mainly as a means of augmenting financial resources, human resources are equally important. International co-operation provides world-wide expertise for optimisation of accelerator and detector design. A broad exchange of ideas and know-how is indispensable to large-scale research in high energy physics.

International co-operation often makes it possible to select physics goals and hence a type of accelerator/detector in an optimal way. For the construction and operation of large facilities, it brings together intellectual, technical, material, and financial resources from different countries. It also allows countries that cannot afford to have their own large facilities to participate. Such participation distributes scientific culture in the world and increases and diversifies the high energy physics community.

Since international co-operation developed rather spontaneously when and where it was needed, various forms of co-operation have evolved for different purposes. Its extent and level ranges from agreements between two institutions to agreements among many governments. There is no particular overall mechanism for co-ordinating these co-operative efforts. Nevertheless, depending on its purpose and scale, each kind plays its role. In some cases, co-operation involves only the exchange of information; in others, it means carrying out projects or research programmes or constructing large facilities in

111

collaboration. The European Laboratory for Particle Physics (CERN), which was built and is managed as a regional institute under a treaty among governments, is an example of long-term co-operation by a large number of countries.

2. The International Union of Pure and Applied Physics

On the researcher side, IUPAP generally seeks better co-operation among the physics societies in many countries. Its guiding principle is that there should be no discrimination of any kind, even when there is political conflict among governments. It has 19 commissions, each of which is concerned with a different field of physical science or with a specific task. Commission 11 (C11) is in charge of high energy physics and created the International Committee for Future Accelerators (ICFA) (see Section 8), which plays an important role in the field.

International conferences

Many conferences, symposia and workshops are organised every year in order to exchange the latest information. Most are open to international participation, so that all interested parties may have access to the new results or ideas. In particular, IUPAP sponsors and the C11 co-ordinates three international conferences, which are held periodically:
- International Conference for High Energy Physics (ICHEP);
- International Symposium on Lepton-Photon Interaction at High Energies;
- International Conference for High Energy Accelerators.

The first two series, which mainly deal with physics results, are held every second year. The third, which addresses accelerator technologies, is held once every three years. The ICHEP series has existed for nearly 50 years and is known as the Rochester conferences, in memory of the university where the conferences were hosted from 1952 until 1957.

These conferences or symposia are hosted on a generally rotating basis by different regions. Although they are often hosted by important research centres or relatively large universities, the scientific programmes are planned through international co-operation.

Visa issues

IUPAP-sponsored conferences have sometimes encountered serious problems concerning visas for participants from countries that are under sanction for one reason or another. IUPAP always asks the relevant government to issue visas to any *bona fide* scientists, so that they may attend the conferences in accordance with the International Council of Scientific Union's (ICSU) principles of free circulation of scientists.

3.   International co-operation in experiments

Discussion of international co-operation in high energy physics can be separated into two areas: experiments and accelerators.

Today, most large experiments are performed as a large collaborative effort by many institutions from many countries. This is regarded as a standard way to build a detector and extract physics results. The number of participating physicists ranges from a few hundred to nearly 1 000, and averages about 400. There are two reasons for these numbers. First, there is a limited number of facilities available, so that physicists from throughout the world are attracted to them, particularly in the case of very high energy colliders. Second, a large number of well-trained physicists are needed in order to solve the many problems involved in building and operating an innovative large detector which can deal properly with the new phenomena produced by these accelerators.

This is a successful outcome of the long tradition of international co-operation in this field. It is now taken for granted that, when a large facility is planned, experiments will be performed with global international collaboration.

If one looks at each experiment in detail, it is evident that the working mechanism is not always the same. Each group has to organise its own activities and take its own decisions. However, in all cases, before a group is formed, interested physicists get together to discuss what they want to investigate, how to approach the objectives, and how to share tasks. Experiment groups are formed through negotiations among many institutions. In order to facilitate this process, workshops or symposia are organised to stimulate the exchange of ideas and to catalyse the formation of groups.

Experiments are usually approved in the following way (see also Chapter 2, Section 4). The host institute, which builds and operates the accelerator, calls for letters of intent or proposals from the broad international community. Each interested group of physicists, usually an existing international group or one formed as described above, responds. The host institute then calls on an advisory committee to examine the proposals. Even when the host institute is a national laboratory, the advisory committee is usually an international body. It examines in detail all aspects of the proposals, in terms of physics, technology, requirements concerning allocation of accelerator time and the beam quality, safety, feasibility, budget, time schedule, requested infrastructure, and so on. The proposals are discussed to some extent in open sessions which anyone can attend, either to listen or to raise questions. Then, using the recommendations of the advisory committee, the host institute makes its selection, finalising the plan in communication with the committee. At the same time, each participating institute requests funding from its financing agency. There is a grey area here, as to which should come first, approval of the proposal or obtaining the budget. The experiment is approved solely on the basis of the physics, feasibility, and the ability of the group; the national origin of the physicists is irrelevant. In this, these decisions respect the ICFA guidelines to be described below.

113

4. How can infrastructures for co-operation on experiments be improved?

When a large group dispersed throughout the world carries out a large-scale experiment, it requires some special infrastructure. A high-quality communications system, for instance, is indispensable. Exchange of detailed drawings or large software packages enables remote institutions to co-operate efficiently, and high-speed fax and high transfer rate computer networking have been very rapidly adopted. However, in some countries in which the telephone network is owned by the government, the available technology is not used as efficiently as it could be.

During the construction phase, many parts of the detector are produced in collaborating countries rather than in the host country and are later shipped to the host institute for installation. Often, partly finished detectors travel around the world to be completed. In some cases, detector parts are tested at a test beam at an accelerator of a third laboratory. This kind of co-operation among laboratories is not exceptional. In such cases, customs procedures or taxation can cause problems and can be costly in terms of time, personnel, and money. Thus, relaxation of the customs procedure for material and tools for basic science will substantially promote international co-operation. The recent disappearance of the COCOM barrier has meant smoother co-operation with the countries of eastern Europe.

Another problem, less frequent but potentially more severe when it occurs, concerns visas for scientists. Experts are frequently needed on the site on short notice in the course of construction or operation. When entering the host country requires obtaining a visa, delays of a few weeks to a few months can occur. It is very important that team members are free to travel to the host country. At the least, a very rapid procedure for issuing visas would improve the situation where that is not possible.

As more and more high energy physicists work in international collaboration, the number of physicists and their families living abroad has increased. Except in a few lucky cases, less attention is paid to their living environment than to that of other groups obliged to work abroad. It may be thought that high energy physicists choose to work abroad for pleasure, and that they can go wherever they would like. When facilities are limited, as they are today, this is not the case. They have little freedom left about where they will work. Often, supporting staff or families are forced to live in difficult conditions, and schooling and health insurance are frequently a problem.

Exchange of scientists is a common practice. In particular, young physicists are encouraged to work abroad for a certain period to gain new experience. The interaction among scientists at the individual level subsequently benefits large-scale international co-operation. In order to promote more frequent exchange of scientists across borders, work permits for them and, where relevant, their spouses should be obtainable as easily as possible. In general, it would be desirable to strengthen exchange programmes.

5.   Bilateral co-operation

Many bilateral co-operation agreements exist at various levels between laboratories, universities, financing agencies, or governments. Agreements made by higher-level bodies usually extend beyond high energy physics, but high energy physics occupies an important place, along with energy or nuclear physics programmes.

Among the bilateral agreements, the US-Japanese Co-operation in Science agreement is one that provides strong support to high energy physics. A five-year agreement, it has been prolonged twice. It provides funding for joint research programmes carried out mainly at facilities in the United States. A substantial share of the funding has recently been used for R&D for detectors and accelerators. The programmes are discussed in a committee composed of physicists and financing agencies from both countries, and the agreement has strengthened co-operative research and understanding. Another remarkable example is the international collaboration for building the HERA collider in Hamburg. It can be viewed as a number of bilateral agreements between DESY and various bodies in different countries. Although the collaborating bodies were not all of the same level and therefore the agreements were not all of the same nature, the collaboration worked very effectively.

Bilateral agreements form a densely woven co-operative network. Although they do not constitute a whole, they operate very well for world-wide co-operation. For instance, broad international collaboration for experiments at Fermilab in the United States or at DESY in Germany is partly supported by research programmes under bilateral agreements. Likewise, agreements between lower-level bodies are helpful in providing channels for the exchange of physicists.

Financial support through bilateral agreements is generally very effective and efficient. Each co-operation programme is of fixed duration, although the agreements themselves may continue and support new projects.

6.   Collaboration with non-OECD countries

Collaboration in high energy physics has in principle been independent of the economic status of the countries. Thus, bilateral co-operation has been common between advanced OECD countries and the former socialist and/or developing countries.

Collaboration with the former Soviet Union

Following rapid changes in the political and economic systems of eastern Europe, the countries of the former Soviet Union (FSU) are undergoing a difficult transition to a market economy. They can ill afford sizeable direct financial contributions to megaprojects or even smaller-scale efforts, but they can still make a contribution. The FSU has many excellent high energy physicists and engineers who can construct parts of accelerators and detectors more cheaply than can be done in the West. Subcontracting subsystems of accelerators or detectors to FSU high energy physics laboratories can save a sizeable fraction of project costs. The difference between the cost in the West and the actual amount of money paid can be considered as a contribution of the FSU laboratories (a "partial compensation" approach). Thus, the FSU laboratories can work as money multipliers.

Collaboration with developing countries

High energy physics laboratories and their facilities are open to all interested scientists, whatever the economic level of their countries. It is desirable to attract potential intellectual resources from everywhere. In the long run, involvement in megascience will support the build-up of science and technology in developing countries. The high energy physics community is well aware of this side benefit to scientific research.

One way of strengthening collaboration with developing countries might be the creation of a distributed laboratory, administered by an international directorate with an international advisory committee, with small branches in different countries. ICFA could play an important role in the creation of such a laboratory. Such an initiative should take into account experience gained in a similar initiative, such as the World Laboratory.

This would be a megaprogramme to be funded by both developed and developing countries. In order to motivate the developing countries to spend money on science, some kind of matching funds procedure could be envisaged. The distributed laboratory could also be used to preserve centres of excellence in high energy physics in eastern Europe.

7. Regional co-operation

International co-operation is less common for building accelerators than for experiments, probably for a number of reasons. Often, a large accelerator is a symbol of prestige or a showcase for the high technology or economic power of a nation or a region. It might indirectly influence defence or energy policy, although there is little chance of this in the near future. It stimulates the economy and leads to further technological innovations. All these aspects tend to make big accelerators a political as well as a scientific issue. This complicates international co-operation. However, there have been successes in this area as well. The most efficient and successful examples are in Europe.

European Laboratory for Particle Physics

The European Laboratory for Particle Physics (CERN) was founded soon after World War II (1954), in order to raise western European particle physics capabilities to the level of those of the United States. The desire to reach this goal was strong enough to overcome difficulties. CERN, located in Geneva, was originally organised as an international body composed of 12 member countries. Each member country contributes in proportion to its net national income, up to a maximum of 25 per cent of the total CERN budget. Decisions are made by the Council, which has two representatives from each member country, one from government, and one from the physics scientific community.

CERN has at present 19 member countries, and its membership has recently increased with the entry of Poland, Hungary, the Czech Republic, and Slovakia.

CERN has constructed a succession of accelerators, from a synchrocyclotron and a proton synchrotron to the latest LEP. It has been the centre of particle physics research in Europe and is now the largest laboratory in the world with respect to budget, personnel, and number of users. It has a relatively fixed budget, a fact that allows it to develop a long-range plan (see Chapter 2, Figure 10). Because it is founded on an agreement among governments, governments of member countries cannot simply and unilaterally change their contributions.

Today, CERN is one of the most productive laboratories and attracts researchers from throughout the world. Most of the physicists working at CERN are visitors from other institutes who remain for a limited period. CERN has its own permanent staff, with a large share of machine physicists and support staff, who have developed various new technologies required by successive research programmes.

JINR at Dubna

JINR, a joint research centre in Dubna, Russia, covers many research fields. When it was founded, it had the proton accelerator with the highest energy in the world. Today, its research is not directed towards the physics of the energy frontier but towards the intermediate energy range, with an emphasis on issues in nuclear physics.

European Committee for Future Accelerators

The European Committee for Future Accelerators (ECFA) was set up in 1963 in order to help co-ordinate the European high energy physics programme. Participating physicists are from member countries of CERN, which itself is considered a ''country'' for the purpose of ECFA representation.

ECFA is an advisory body to CERN management, the CERN Council and its committees, DESY management and its Scientific Council, and various other organisations, both national and international. ECFA consists of Plenary ECFA, Restricted ECFA, Chairman, Secretary, and permanent or *ad hoc* working groups. Plenary ECFA normally holds two meetings a year; these are usually public. Restricted ECFA is composed of one member per country. It assists and advises the Chairman and the Secretary and acts as the channel of communication to the physics community and national institutes and authorities in each participating country. The CERN and DESY Directors-General are *ex officio* members. All major facilities constructed in Europe are discussed in ECFA. Special workshops are organised to investigate the details of the projects, so as to obtain a regional consensus. For instance, ECFA supported the development of both LEP and HERA in this way.

The guidelines of ECFA define its aims and activities:

Aims:

"Long-range planning of European high-energy facilities – accelerators, large-scale experimental equipment and computers – adequate for the conduct of a valid high-energy research programme by the community of physicists in the participating countries and matched to the size of this community and to the resources which can be put at the disposal of high-energy physics by society. Duplication of similar accelerators should be avoided and international collaboration for the creation of these facilities should be encouraged if essential and efficient for attaining the purpose.

"Equilibrium between the roles of international and national laboratories and university institutes in this research, and a close relation between research and education in high-energy physics and other fields. Adequate conditions for research and a just and equitable sharing of facilities between physicists, irrespective of nationality and origin, as conducive to a successful collaboration effort."

Activities:

"To achieve these aims ECFA can engage in – among others – the following activities:

a) regular meetings of Restricted and Plenary ECFA;

b) ad hoc symposia and conferences sponsored or organised by ECFA;

c) study groups, set up by ECFA, or jointly with other organisations, for special problems;

d) demographic studies of the high-energy physics community and resources in the ECFA countries, repeated at regular intervals."

8. Global co-operation

So far, high energy physicists have not built any facility in inter-regional co-operation. HERA is the only instance in which co-operation was partly obtained from another region. The project was proposed by DESY (Hamburg) and strongly supported by ECFA as a German project with international co-operation. Through negotiations between DESY and other laboratories or authorities prior to the decision to go ahead with the project, six nations joined the construction by making in-kind contributions. Most of the participating nations are European but two, Canada and Israel, are not. In addition, Poland and China co-operated by providing personnel. Co-operation, including technology transfer between DESY and the collaborating institutes, went well. The responsibility for the final commissioning of the collider and its operation remained with DESY. This scheme, now called the "HERA model", was a success and may be used again for another construction project, depending on the size of the facility and the willingness of other countries to engage in this kind of co-operation.

International Committee for Future Accelerators

Global co-operation was first discussed in the 1970s for a future very big accelerator complex, and the discussions led to the establishment of ICFA by IUPAP-C11 in 1977. ICFA members are selected regionally and approved by IUPAP-C11. Reflecting the political structure of the world when it was set up, ICFA has 14 members: the United States (3), Western Europe (3), Russia (3), JINR member countries except Russia (3), Japan (1), China (1), a fourth region (1), and the chairman of IUPAP-C11 *ex officio*. ICFA normally meets once or twice a year.

The terms of reference of ICFA, as decided by C11, are:

"to organise workshops for the study of problems related to an international super energy accelerator complex (VBA) and to elaborate the framework of its construction and of its use;

"to organise meetings for the exchange of information on future plans of regional facilities and for the formulation of advice on joint studies and uses."

Although ICFA's members did not explicitly aim to build a big facility, it was hoped that if a scheme for global co-operation was developed, a big accelerator might be realised.

While discussing the technologies of future accelerators, ICFA, at its fifth meeting at CERN in 1980, formulated guidelines for the use of major regional facilities for high-energy particle physics research. These guidelines have since been respected by all laboratories in the field, both national and international.

ICFA's motivation was stated as follows:

"Considering that in the future major experimental facilities for high energy particle physics, notably the very largest particle accelerators and colliding beam machines, are likely to be few in number, probably only one of each type of the very highest energy and that these machines will be located in different regions of the world;

"And recognising that experimental physicists from all regions will wish to gain access to these few machines in order to pursue their research;

"ICFA proposes that the regional laboratories operating these facilities should adopt a common policy towards experimental physicists from other regions seeking to use the facilities they operate."

The guidelines proposed are as follows:

"1. The selection of experiments and the priority given to them are the responsibility of the laboratory operating the regional facility.

"2. The criteria used in selecting experiments and determining their priority are:

a) scientific merit;

b) technical feasibility;

c) capability of the experimental group;

d) availability of the resources required.

"3. It is expected that teams from other regions will normally wish to join with local regional teams to form experimental groups in proposing and carrying out experiments using a regional facility. The national or institutional affiliations of the teams should not influence the selection of an experiment nor the priority accorded to it.

"4. The availability of the resources needed for the experiment are examined at the time of selection of the experiment. The contribution of each team and of the Operating Laboratory to an experiment are the subject of agreements drawn up between the Operating Laboratory and the authorised leaders of the teams in the experimental group. When appropriate, realisation of the proposal approved may be effected within the framework of bilateral and multilateral agreements in force or newly reached agreements.

"5. Operating Laboratories should not require experimental groups to contribute to the running costs of the accelerators or colliding beam machines nor to the operating costs of their associated experimental areas.

"6. It is expected that averaged over a reasonable period of time the application of guideline 2 above will lead to a balanced use of the major new facilities by the regions concerned. However, if at any time an Operating Laboratory finds that the participation of teams from other regions in their experimental programme is becoming excessive, the Operating Laboratory may be obliged to limit that participation. Any such action should be accompanied by discussions with the relevant authorities of the regions concerned and consultations with the other operations laboratories subscribing to the Guidelines laid down in this document."

Modification of the ICFA policy

The advanced technology for accelerators, which was the focus of ICFA's concerns, was incorporated in proposals for the Accelerating and Storage Complex (UNK) in the former Soviet Union and the Superconducting Super Collider (SSC) in the United States. In 1984, when the US government decided to build the SSC as a national project, ICFA modified its policy after serious discussions.

It issued the following revised guidelines in 1985:

"to promote international collaboration in all phases of the construction and exploitation of very high energy accelerators;

"to organise regularly world-inclusive meetings for exchange of information on future plans for regional facilities and for the formulation of a consensus on joint studies and uses;

"to organise workshops for the study of problems related to super high-energy accelerator complexes and their international exploitation".

At the same time four standing panels were formed to study problems of accelerator and detector technologies. Each panel organises workshops or courses in its area on the basis of world-wide co-operation. Every three years, ICFA organises other general seminars on future perspectives in high energy physics to discuss the latest developments in high energy physics and accelerator technology, as well as international co-operation. The first ICFA seminar was held at KEK (1984); it was followed by those at BNL (1987), Protvino (1990), and DESY (1993).

Because of its very idealistic original aims and its later retreat, ICFA is sometimes regarded as having no real power. However, this view may be short-sighted. Although its members include directors of major laboratories and professors of leading institutes in the field, and in a sense represent users, it is not ICFA's role to exert any control over or co-ordination of the development of research. Rather, ICFA's aim is to facilitate communication and understanding among the world's high energy physicists. There is no other place for laboratory directors to meet around a table and engage in a frank exchange of views. For the sound development of the field, this is very important.

It is widely believed in the high energy physics community that free competition is essential and that it is a driving force of advancement. Like international co-operation, competition has been effective in ensuring, world-wide, a high level of activity, optimum use of resources, and a variety of research. It has also helped in the distribution of large projects over the world. Thus, while the freedom of others is accepted, once co-operation agreements are made, they are respected. ICFA's guidelines for the use of facilities offer a good example, and, in a sense, ICFA has a certain power.

ICFA's reaction to the cancellation of the SSC

The joint meetings of ICFA and laboratory directors following the demise of the SSC project (October 1993) offer a good example of the close communication that often leads to co-operation. A special meeting held in Geneva in December 1993 was followed by a second in Vancouver in January 1994. At the end of a two-day meeting, a statement was issued on 17 January 1994. It read as follows:

"Statement of the International Committee for Future Accelerators (ICFA) on International Collaboration in the Construction of Major High-Energy facilities.

"High-energy physics seeks to discover basic principles that underlie the working of the physical universe through the exploration of the building blocks of matter and the forces among them. World-wide effort over the past half-century has produced a remarkably successful theoretical picture describing all matter and energy as built of certain constituents, interacting through specific forces according to general principles of symmetry, relativity and quantum mechanics. Yet the picture contains gaps – profound questions that can only be answered with new facilities. The answers to these questions hold the promise of yielding an historic unification of ideas and principles, as significant as those that have marked past revolutionary advances in scientific understanding.

"Particle accelerators and detectors have served as experimenters' most successful tools for this exploration of the subatomic world, and will do so in the foreseeable future. To probe matter and energy at the point where revolutionary discoveries are expected, particle accelerators of energies higher than are now available must be built. Drawn by the importance and the scientific challenge of such discoveries, high-energy physics experimenters have traditionally pooled their resources to

121

build detectors across international boundaries, forming large regional centres and scientific collaborations to meet the higher costs of advancing exploitation.

"The termination of the Superconducting Super Collider Project, the highest energy collider ever begun, is a very great loss to the world high-energy physics community. The outcome illustrates the need to make the construction of new large facilities the result of a world-wide strategy, in the same collaborative spirit that has characterised the construction of major experimental detectors.

"Following the cancellation of the SSC the Large Hadron Collider (LHC) at CERN now offers the only realistic opportunity to study multi-TeV hadron collisions. ICFA notes that the LHC project is now ready for approval and is currently being evaluated by the CERN Council. The energy and luminosity of the LHC represent a great advance over Tevatron, now the highest energy collider in operation (seven times in energy and a thousand times in luminosity). There are compelling arguments that fundamental new physics will appear in the energy domain that will be opened up by the LHC, including the origin of electroweak symmetry breaking (and hence the origin of mass). The LHC will remain a unique facility for the foreseeable future and ICFA considers that it is now the correct next step for particle physics at the high-energy frontier. ICFA therefore hopes that the nineteen Member States of CERN will quickly approve the LHC for timely completion. ICFA notes the world-wide interest in participation in the project. ICFA urges that appropriate mechanisms and means be found to allow this to happen and that the LHC be available for research by the world particle physics community.

"In the not-too-distant future, accelerator specialists will complete the research and development necessary to begin the design of an electron-positron collider capable of exploring the comparable mass region. As has been the case in the past such an approach will be complementary to what will be done with proton-proton colliders. ICFA notes that research and development for the design of a large electron-positron linear collider is being carried out under an international collaboration. The signatories of the memorandum of understanding among the participants of that collaboration have pledged to admit all institutions that are prepared to make significant contributions to the research and development effort. The participants further share a common vision of a facility that will be built as a world-wide collaboration. ICFA continues to strongly endorse the goals of this collaboration.

"ICFA believes that the time has come for the governments of all nations engaged in the science of high-energy physics to join in the construction of major high-energy facilities, so that this unique endeavour can continue to go forward."

Reinforcement of ICFA

When ICFA reaches an agreement, as in the case described above, ICFA members for each region and/or laboratory directors transmit the decision to the high energy physics community and to the relevant governments. Since there is no official link between ICFA and governments, this is done unofficially and, in most cases, works well. Laboratory directors, in particular, have close links to their governments. In Europe, the process is accomplished via ECFA, which acts as an interface between ICFA and the European high energy physics communities and administrations.

However, the situation is still somewhat unsatisfactory. At present, ICFA recommendations are transmitted differently in each region. In some cases, the procedure is not very transparent, even ill-defined. In particular, in countries without big laboratories, which thus lack strong communities in the field, communication with ICFA is weak.

In order to stimulate better international co-operation, it may be desirable to reinforce ICFA by strengthening its link to the user community. It is likely that a larger representation from the user community, as is the case for ECFA, would help.

Following its reinforcement, detailed discussions in ICFA and in IUPAP-C11 might lead to an expansion of ICFA's role. In the future, it may be appropriate to establish a mechanism for formal communication with the financing and decision-making bodies at government level through, or based on, ICFA. Chapter 6, Section 7, further elaborates on this possibility.

9. Summary

International co-operation has a long-standing tradition in high energy physics. The field has advanced in a spirit of complete openness, through scientific exchanges and sharing of facilities.

International co-operation often leads to an optimum way of selecting physics goals. It brings together intellectual, technical, material, and financial resources from different countries for the construction and operation of large facilities and/or experiments. In this way, high energy physics has been very successful in generating intellectual achievements and in promoting international understanding.

Many levels of co-operation have been developed between various bodies for different purposes. Each type plays its role. On the research side, IUPAP plays the central role. IUPAP's C11 is charged with the promotion of international co-operation in the field of high energy physics. Many bilateral agreements between governments and other financing bodies all over the world provide financial support for co-operation.

It is common practice for an experiment at a large facility to be performed through international co-operation, whether the facility is national or international. The procedure of setting up a group is well established although the organisation of each particular group varies widely.

There are two regional centres in Europe, CERN and JINR, where big facilities have been constructed through international co-operation. CERN has been very successful in building machines that have played a leading role in high energy physics. So far, HERA, built at DESY, a project for which there were remarkable in-kind contributions from abroad, including two non-European regions, is the only accelerator constructed through global co-operation.

In order to promote international co-operation, it is desirable to eliminate, as much as possible, practical barriers raised by visa applications, customs clearance, and import duties, as they often cause difficulties that result in a waste of time and financial resources.

There are bodies within the high energy physics community that seek to promote international co-operation. In Europe, ECFA has been established to help co-ordinate European activities by providing advice to laboratory management and by organising seminars. Globally, ICFA was set up by IUPAP-C11 to study the problems of very large accelerators that might be built and used in world-wide co-operation. While studying the technology and the plans for such a machine, ICFA established guidelines for the international use of major regional facilities. These are respected by all laboratories in the field.

ICFA serves to promote communication and understanding among high energy physicists in the world. There is still some way to go to achieve its original aim, and it might be useful to reinforce the organisation. One possibility could be to strengthen ICFA's link to the user community, after which its role with respect to governments might be expanded.

Chapter 5

New Projects and Long-term Developments

1. Introduction

In the past two decades, high energy physics has, to a very large extent, been driven by experiments at storage rings (colliders). The most prominent advances were the discovery of the charmed quark, the heavy lepton tau, and the carriers of the weak and strong forces, the heavy bosons W and Z, and the gluon g. In fact, a large part of our knowledge of matter and forces, known today as the Standard Model, stems from collider experiments.

It is therefore not surprising that the main thrust of further developments of facilities for future research in high energy physics involves more powerful colliders. Improved machines can basically be achieved either through higher energy or through higher luminosity. Table 10 summarises the present situation.

Table 10.　**Present and future colliders**

	High energy			Medium energy
	e^+e^-	ep	$pp + p\bar{p}$	e^+e^-
a)	LEP SLC TRISTAN	HERA	Tevatron (Sp$\bar{p}$S)	CESR (DORIS) BEPC
	⇓	⇓	⇓	⇓
b) Beyond 2000	Energy frontier = higher energy			Precision frontier = higher luminosity
	⇓	⇓	⇓	⇓
	e^+e^- Linear collider	LHC × LEP Linac × LHC	LHC UNK	Factories φ: DAΦNE, Novosibirsk c, τ : ? B: SLAC, KEK

a) Existing colliders. Machines in brackets have recently been closed down.
b) Future colliders.
Source: Author.

High energy luminosity is pursued in so-called factories, which are all e⁺e⁻ colliders. However, the discussion will be restricted to the colliders with more than 10 GeV centre-of-mass energy, which are all of the B factory type. The two lines of research currently being followed are pp and e⁺e⁻ colliders, and an ep option is being discussed at CERN for the LEP/LHC complex. Figure 18 illustrates the rapid progress in the field of high energy colliders; it is often referred to as the Livingston plot after the accelerator pioneer who first used it. It shows the exponential development of available centre-of-mass energy over the past decades. This chapter will follow the classification indicated in Table 10 and is organised by:

- B meson factories;
- large circular proton colliders;
- electron-positron linear colliders (EPLC).

2. B meson factories

General considerations

The b quark appears to offer unique ways to study some of the most interesting questions in particle physics today. Those advantages stem partly from the fact that it is heavy and partly because all of its decays are somewhat inhibited, so that it lives a relatively long time, about 1.5×10^{-12} s. The three areas where study of b quarks appears to offer unique opportunities are:

- Study of the top quark.
- Study of CP violation, the asymmetrical behaviour of matter and antimatter. In the b quark system, this is reflected in different decay rates or different temporal evolution for the b quark and its antiquark.
- Existence and properties of much more massive states, such as the Higgs boson. This question can be studied by looking at certain rare decays of B mesons, states containing a b quark as one of its constituents.
- Determination of some of the basic parameters of the Standard Model, *e.g.* elements of the weak mixing matrix. This information can be obtained from the systematic study of several different B meson decay modes.

Meaningful studies of all these topics require copious sources of b quarks and high-precision detectors to suppress backgrounds. To answer several of the questions in these areas, about $10^7 - 10^8$ b quark states will have to be analysed. This is one of the reasons for upgrading existing b quark production facilities or for building new ones.

CP violation study requires very high B production rates, but it also makes additional demands. The decay channels that are expected to exhibit large CP violation effects have very low branching ratios. Very roughly, about 10^8 analysable b quarks are needed to obtain significant results on CP violation. Such data samples should be available in next generation of hadron machines, such as the LHC. Because of the high rates at these machines, the experimental challenges are formidable. At hadronic colliders, B produc-

Figure 18. The expanding energy frontier of colliders

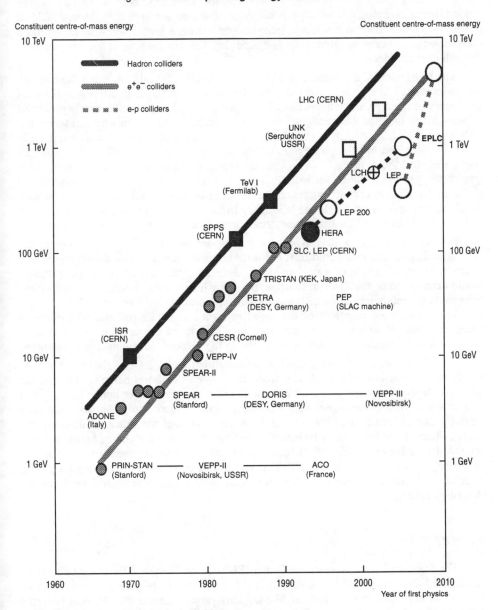

Note: "Livingston plot" of pp, ep, and e⁺e⁻ colliders, including new projects. The constituent centre-of-mass energy is plotted as a function of the first year of operation.
Source: Author.

tion constitutes less than 1 per cent of total events. Therefore, because of large background, many potentially interesting B decay channels do not appear to be amenable to precise analysis at hadron machines. Hence, there has been great interest in exploiting a complementary means of producing and studying CP violation in B mesons, namely through e^+e^- colliders.

The most certain, and most complete, way to study CP violation using the b quark system produced in e^+e^- collisions also requires measuring the time difference between the decays of b and $\bar{b}$, *i.e.* b quarks and b antiquarks, which are simultaneously produced in pairs. Since the time difference in this domain can be measured only by measuring the difference in path lengths, it is important that the b quark is produced with a relatively large velocity. The most copious rate for $b\bar{b}$ quark production in collisions occurs just above their threshold, at the $\Upsilon(4S)$ resonance. Thus, in the centre-of-mass system, the b quarks are produced almost at rest. In the e^+e^- colliders built to date, electrons and positrons circulate in opposite directions in the same beam pipe and collide head on with equal and opposite momenta. In the laboratory frame, the b quarks are also almost at rest, and it is not possible to measure their decay times from their path lengths.

Several years ago, a new type of e^+e^- collider was proposed to overcome this difficulty. The two beams would travel in different beam pipes and could therefore have different energies. These energies could be adjusted so that the centre-of-mass energy would correspond to the $\Upsilon(4S)$ resonance, but the whole system would move with large velocity in the laboratory system. The B mesons produced would have substantial laboratory velocities (more than half of the speed of light), and their path length difference could be measured with sufficient accuracy. Such colliders are referred to as asymmetric colliders and, if optimised for operation at the $\Upsilon(4S)$ resonance, asymmetric B factories.

To achieve sensitivity sufficient for meaningful studies of CP violation, about 10^8 B meson pairs need to be studied. Therefore, the new e^+e^- colliders being constructed not only have to operate in the asymmetric mode but also have to achieve luminosities almost two orders of magnitude higher than what is available in existing machines. Two such facilities are under construction: one at KEK in Japan and the other at SLAC in the United States. In addition, the CESR facility at Cornell, also in the United States, is being upgraded to achieve significantly higher luminosities. It will operate in the symmetric mode, so, unlike the other two facilities, it will not be capable of performing a comprehensive study of CP violation. Plans for each of the two asymmetric facilities are described below.

B factory at SLAC

The Stanford Linear Accelerator Center (SLAC), the Lawrence Berkeley Laboratory (LBL), and the Lawrence Livermore National Laboratory (LLNL) have begun construction of an asymmetric B factory at SLAC. This project, called PEP-II, is a two-ring upgrade of the existing one-ring PEP facility. In PEP-II, the high energy ring will store 9 GeV electrons that will collide with a second ring of 3.1 GeV positrons. The high luminosity is obtained with large beam currents, about 1 A in each beam. To reduce single-bunch instabilities, each beam is distributed among 1 650 bunches. The SLAC site provides several advantages for the construction of an asymmetric B factory. No conventional construction is required. The 2.2 km PEP tunnel is fully adequate for accommodating the two rings; no additional power or utilities are required. Fully 80 per cent of the present PEP facility is reusable for PEP-II. The PEP magnets are reused for the 9 GeV ring and components from the present vacuum system will be used for the 3.1 GeV ring. Both rings require a new (common) RF system and largely new vacuum systems. The powerful existing SLAC linear accelerator (linac) provides low-emittance, high intensity beams for injection. Dedicated, on-energy injection is being provided by new by-pass lines housed in the linac tunnel. The availability of the PEP tunnel, components, and infrastructure, along with the absence of a need for conventional construction make the upgrade relatively inexpensive ($177 million) and allows rapid implementation. First collisions are anticipated at the end of 1998. The Institute for Nuclear Science in Novosibirsk and the Institute for High Energy Physics in Beijing are expected to join as collaborators to build the machine.

The detector for PEP-II will comprise a fully international collaboration of 400 to 500 physicists. Currently in the design phase, it will look much like a conventional 10 GeV electron positron colliding beam detector, with a few wrinkles. Collisions at PEP-II occur every four nanoseconds, with a physics event rate of about 10^3 Hz. As a result, the data acquisition system will resemble that of a proton detector. The asymmetry of the collisions puts a higher than normal premium on "forward" detection.

The individual subsystems of the detector can be implemented using conventional techniques except for particle identification. The vertex detector will use silicon strip technology, charged-particle tracking will be accomplished with a 1-1.5 Tesla magnet and a large cylindrical drift chamber, photon and electron detection will be done with a cesium iodide (CsI) calorimeter, and muons will be selected on the basis of range in an instrumented iron absorber which doubles as the magnet flux return. For particle identification, pion/kaon separation is needed up to (high) momenta of 4 GeV in a device which must be less than 15 per cent of radiation length in thickness. The choice of this device awaits the outcome of ongoing R&D; all possibilities employ Cerenkov radiation.

The detector collaboration currently involves physicists from the United States, Canada, France, Germany, the United Kingdom, Italy, Russia, and China. CP violation is the primary goal of the B factory. However, the very high luminosity provides an excellent laboratory for studying b quark, c quark, tau lepton, and two-photon physics. At the minimum, a ten-year programme of studies is foreseen. The B factory will be a valuable training ground for young physicists; it is anticipated that upwards of 100 students will obtain PhDs from the B factory programme.

B factory at KEK

The KEK Laboratory has proposed to build an asymmetric B factory as a second phase of the TRISTAN project. The project is in the government budget plan for JFY 1994, and construction is scheduled to begin soon. The first beam collisions are expected in 1998.

Both electron and positron rings will be installed in the TRISTAN main tunnel; the nominal energy of the electron beam will be 8 GeV and that of the positron beam, 3.5 GeV. There will be a capability for having collisions in two places. Maximum luminosity of 2.2×10^{33} cm^{-2} s^{-1} can be achieved with beam collisions at only one interaction point. In the initial configuration the beams will collide head on.

The original TRISTAN injection system will be modified to provide the beams of required energy and intensity. The present 2.5 GeV linac will be upgraded to 5 GeV. The existing 5 GeV electrons will be injected into the accumulator ring in the TRISTAN tunnel. In the positron injection mode, a metal target will be inserted at the 1.5 GeV point of the linac. The produced positrons will be accelerated to 3.5 GeV in the latter part of the linac and then injected directly into the positron storage ring.

The beam currents required to achieve the design luminosity are 0.52 A for electrons and 0.22 A for positrons. Each beam is divided into 1 024 bunches. There are plans to increase the luminosity later by going to a finite crossing angle; this would allow an increase in the number of bunches without exceeding the beam-beam interaction limit. The beam parameters described above require a new RF and vacuum systems to achieve sufficiently long beam lifetimes (warm temperature and superconducting cavities are being developed for the RF system). The cavity structure is being designed so that the induced higher mode oscillations are damped sufficiently before the arrival of the next bunch. A new copper vacuum system is being developed to handle the high levels of synchrotron radiation. The bunch size of the beams at the interaction points is expected to be 1 cm in length, 3 μm in height, and 300 μm in width. A wiggler magnet system will control the beam properties.

The laboratory has recently put out a call for letters of intent for experimental programmes. All interested physicists from any country are invited to participate. At the present time, a large international collaboration, including scientists from Japan, China, Korea, Taiwan, Russia, and the United States, has come together to initiate the preliminary design for the detector. Some of the features of the detector currently under consideration are: good tracking capability that allows reconstruction of B vertices with precision of 70 μm, good particle identification achieved with time-of-flight counters and ring-imaging Cerenkov counters, and a CsI electromagnetic calorimeter. A pipeline trigger and a computing system with many parallel processors are planned, in order to handle the expected data rates of 100 megabytes per second.

3. Large circular proton colliders

General considerations

As pointed out in Chapter 1, one of the most burning questions in particle physics concerns the generation of mass. The mechanism preferred by the Standard Model would manifest itself by the existence of Higgs particles. The mass of these particles is essentially unknown; however, consistency arguments indicate that at an energy scale of no more than a few TeV, either the Higgs particles or some other new phenomena should appear.

To penetrate into this energy regime requires colliders that produce collisions of this energy at the level of elementary constituents such as leptons or quarks. They might be either e^+e^- colliders of a few TeV or pp colliders of a few tens of TeV. The higher energy is needed for pp colliders because protons are aggregates of quarks and gluons, each of which carries only part of the proton momentum.

Besides this main motive, and to mention only the most prominent examples, pp colliders of this energy also provide extremely powerful instruments to:

- study B meson physics, including CP violation;
- search for new particles, including supersymmetric particles;
- perform heavy ion physics if the rings are filled with heavy ions instead of protons.

Past history has indicated that the secrets of nature are often unpredictable. Whenever a new energy regime has become available, important and often surprising discoveries have followed. High energy pp colliders are excellent exploratory tools, and it is likely that such a machine would also make available a wealth of new information.

All the hadron colliders in the world have been built according to the same principle, first used at the Intersecting Storage Ring (ISR) at CERN; two counter-rotating proton (or antiproton and proton) beams are accelerated in circular synchrotron structures and collided at up to eight collision points. Detectors are built around those points to detect the debris of the violent collision processes. Either two proton beams or one proton and one antiproton beam are used.

In the latter case, a single synchrotron structure with only one vacuum pipe is able to accelerate and contain both beams. This method, which is very economical and has been successfully applied at the Spp̄S at CERN and the Tevatron at Fermilab, cannot be used for the next step hadron collider, since the beam current required to match the falling cross-section cannot be delivered efficiently by antiproton sources. The two required counter-circulating high current beams must be proton beams, and therefore cannot be kept in the same vacuum pipe. Thus, both projects for future pp colliders, the SSC and the LHC, were based on two proton beams circulating in two different synchrotron structures.

In Russia, the UNK project is presently under construction in Serpukhov. The tunnel for this machine has been completed and, as a first stage, a 600 GeV proton accelerator is being assembled. The schedule for completing the ultimately planned 3 TeV accelerator is presently unclear. The UNK complex can play an important role in pursuing many of the fixed target programmes, which will be partly closed down or reduced elsewhere because of new projects such as the LHC at CERN.

Following the cancellation of the SSC – construction of which had already started – the LHC at CERN is now generally considered to be the only realistic opportunity for studying multi-TeV hadron collisions in the near future.

The Large Hadron Collider project

In 1987, after extensive discussions, notably at the La Thuile workshop, the CERN management decided to pursue a pp collider, the LHC, in the LEP tunnel. After studies of optimisation, the CERN management came forward with the definitive proposal for the LHC, which will have two options (see Table 11) and a later possibility for ep collisions.

Table 11. **Main parameters of the Large Hadron Collider (LHC)**
(June 1993)

Collisions		pp	Lead ions
Operational energy	TeV	14	1 148
Dipole field (max.)	T	8.65 (9.0)	
Luminosity	$cm^{-2} s^{-1}$	10^{34}	10^{27}
Luminosity at beam-beam limit		$(2.5 \ 10^{34})$	
Bunch interval	m/ns	7.5/25	40.5/135
Particles per bunch		10^{11}	10^{8}
Particles per beam		$2.8 \ 10^{14}$	$4.7 \ 10^{10}$
Number of experiments		2	1

Source: European Laboratory for Particle Physics (CERN).

The basic idea, which gives the project its particular advantage in terms of cost-saving, is to build it in the existing LEP tunnel. Figure 19 shows an artist's view of how the new LHC magnets will be installed on top of the LEP machine. In addition, the CERN site and all its infrastructure can be used, as shown in Figure 20. The injection complex will include most of the existing CERN machines which will, once they have been adapted, provide the LHC with protons, ions and electrons.

Of course, the price paid for taking advantage of the existing tunnel is a limitation on the attainable energy. Given the constraint of the circumference of the LEP (27 km), the main free energy parameter (except for details in the machine lattice) is the dipole field strength of the main bending magnets. Consequently, the dipoles are the important and decisive elements of the machine. Figure 21 shows a sectional drawing of a super-conducting dipole magnet. For reasons of cost-effectiveness and of the space available in the tunnel, the so-called "two-in-one" solution – in which the beams rotate in opposite directions in two separate vacuum pipes and in two superconducting magnets in one common iron yoke – has been chosen.

More than 1 000 of these dipole magnets, each about 13 m to 15 m long will be needed to equip the length of the LHC. To reach the highest possible field, the magnet coils have to be superconducting. The design of the magnet and the production tooling are still being finalised.

LHC magnet coils are made of copper-clad niobium-titanium cables. This technology, invented in the 1960s at the Rutherford-Appleton Laboratory (United Kingdom),

132

Figure 19. **LHC in the LEP tunnel**

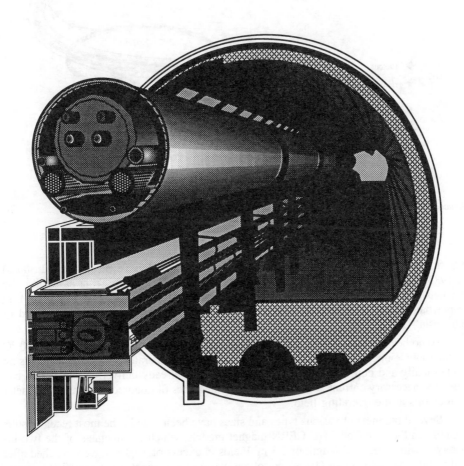

Source: CERN (Geneva).

133

Figure 20. **The CERN machine complex including LHC**

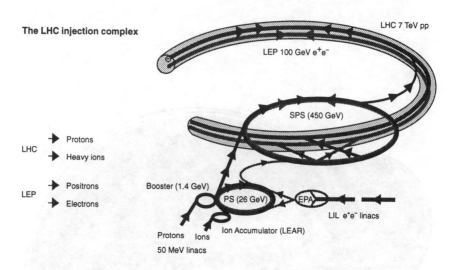

Source: CERN (Geneva).

was first used in a superconducting accelerator at the Fermilab Tevatron (United States) in 1983. The Tevatron magnets currently run at peak fields of 4.5 Tesla at 4.2 K. The magnets in the electron-proton collider HERA, at the DESY Laboratory (Germany) go somewhat higher, to around 5.5 Tesla. The SSC prototype magnets reached a field of 6.6 Tesla.

To go beyond these levels, LHC magnets will be operated at 1.9 K, or almost 300 °C below room temperature. This unusually low operating temperature puts new demands on cable quality and coil assembly. European industry is already delivering cables that can carry the necessary current at 1.9 K and withstand forces of hundreds of tonnes per meter in the coils at the operating field.

Several magnets of various types and sizes have been tested. The most recent results come from the so-called LHC CERN magnet models, which are magnets of the two-in-one type with a reduced length of 1.3 m. Fields in excess of 10 Tesla were reached after full testing cycles. A large (10 m) prototype two-in-one magnet using the HERA super-conducting cable has been tested at Saclay. The maximum field reached was 8.25 Tesla but was limited by the amount of superconducting cable in the magnet.

In view of the lattice optimisation and the experience gained with the test magnets, the operational parameters have been slightly modified with respect to the original design (Table 11). The operational energy will be 14 TeV at a field strength of 8.65 Tesla. This should leave enough margin on the field strength for the safe and stable operation of the LHC.

Figure 21. **Cross-section of an LHC dipole magnet**

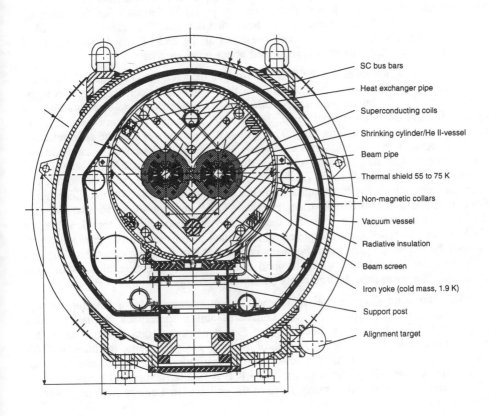

SC bus bars

Heat exchanger pipe

Superconducting coils

Shrinking cylinder/He II-vessel

Beam pipe

Thermal shield 55 to 75 K

Non-magnetic collars

Vacuum vessel

Radiative insulation

Beam screen

Iron yoke (cold mass, 1.9 K)

Support post

Alignment target

Source: CERN (Geneva).

To produce the necessary rates of elementary processes of interest, the somewhat limited energy potential of the LHC has to be compensated by higher luminosity. The projected luminosity of 10^{34} cm^{-2} s^{-1} at a bunch spacing of 25 ns (previous designs had even foreseen shorter bunch spacings) is a major challenge, more to the experiments than to the machine itself.

At first, the LHC will be used to investigate proton-proton collisions at a total energy of 14 TeV, but collisions of heavy ions are also being envisaged, in order to study the behaviour of nuclear matter at energies currently not achievable (150 TeV for impacts between lead ions). Electron-proton collisions could also be achieved at a later stage, using protons from the LHC and electrons from the LEP. Clearly, an extremely rich scientific programme is possible.

Out of the eight intersecting points of the LHC, two interaction areas will be used for dumping and cleaning the beam. Two areas are presently foreseen for the major pp experiments and one for a general purpose heavy-ion detector. The calendar for the LHC and the LEP plans for LEP-II to cease operation at the end of this decade and LHC (machine and detectors) to be commissioned about two years later.

For the first proton-proton collider, two major detectors, ATLAS and CMS, have passed the first hurdle, which is acceptance of their letters of intent; the next step is the preparation of the technical proposals. While both are general purpose detectors that can be used for a broad physics programme, the technical solutions chosen are quite different and in many respects complementary. This will ensure comprehensive and safe detection of all aspects of the collision process. The CMS design is based on a conventional selenoidal configuration, but with very large dimensions (14 m long, 5.9 m diameter bore solenoid, 4 Tesla field); the ATLAS detector, instead, has a small inner solenoid and a huge air core toroid for muon measurements. Both detectors aim at excellent muon detection but use quite different strategies. The ATLAS detector allows for two independent muon measurements in the inner detector and the outer air core toroid, while the CMS strategy is based on a complete and redundant reconstruction of the muon tracks through the whole detector from the beam vertex to the outer muon chambers. The two calorimeter designs are well suited to good electron/photon detection, along with good capabilities for measuring hadronic jets and missing energy.

The above pp experimental programme is complemented by a detector facility, ALICE, designed to study heavy ion collisions, and a specialised b quark physics programme which has not yet been decided. Smaller, more dedicated experiments are expected to follow at later stages. On each of the large general purpose facilities, ATLAS and CMS, groups of about 800 physicists from 90 institutions work together in a world-wide collaboration. The group working on ALICE is smaller, about 230 physicists from 42 institutions. In all groups a major share of the collaboration comes from non-member countries of CERN.

The cost of building the machine, including the installation of the experimental area, is estimated to be about SF 2.5 billion, excluding the salaries of CERN personnel. Maximising the use of all the existing infrastructure (injectors, tunnel) will result in substantial savings. The initial cost of the two major multipurpose detectors now under preparation (ATLAS and CMS) will be about SF 350 million each, with construction taking place in stages; the final cost will be over SF 400 million. These last figures, established according to the methods used in Europe, do not take into account the salaries of staff belonging to the institutes involved in the collaboration.

Assuming that the annual resources of CERN provided by member countries remain at the currently agreed level, and that the project is funded over the period 1995-2005, additional contributions of at least SF 500 million have to be found from inside or outside member countries. Some forms of association could be devised so that non-member countries that would like to be significantly involved in the construction and operation of the machine could contribute to CERN's budget and express their views in its official bodies. Bilateral agreements could then be concluded between CERN and the "associate countries"; it would probably be equitable for the budget contributions of these associate countries to be calculated as a function of the extent to which they use the installations, measured, for example, in terms of the number of scientists expected to take part in the experiments.

4. Electron-positron linear colliders

General considerations

As outlined above, e⁺e⁻ collisions at a few TeV would present an equally powerful tool for studying the energy frontier. In various ways, e⁺e⁻ and pp collisions are complementary approaches. History shows that, in many cases – for example, the b and t quarks and the heavy vector bosons – discoveries were made at p accelerators or colliders, while detailed studies were carried out at e⁺e⁻ colliders. These studies have yielded a wealth of information on the basic constituents of matter and the forces that act between them. This is due to the inherent simplicity of the annihilation process in e⁺e⁻ collisions, where the interaction takes place directly between basic constituents. Detailed extrapolations show that the high visibility, the well-defined event topology, and the excellent signal-to-background ratio that characterise e⁺e⁻ collisions at present energies will persist in the energy range opened up by future linear colliders.

To go beyond the energy of the LEP-II in electron-positron collisions, namely about 200 GeV, the next generation of colliders must be linear machines, in order to avoid what would be prohibitive energy losses from the beams by synchrotron radiation in circular machines. Building these installations means finding answers to very difficult technical problems, but the successes achieved by the SLAC at Stanford, the world's first e⁺e⁻ linear collider (EPLC) in service, with an energy of 90 GeV, are encouraging.

Various recent workshops on the potential for physics of the e⁺e⁻ linear colliders in the centre-of-mass energy range of several hundred GeV seem to show that such a machine may be able to provide important complementary information to a hadron machine like the LHC. For example, the measured value of the top quark mass (174 ± 20 GeV/c²) and the tendency of the supersymmetric models to predict rather low Higgs masses make even a 500 GeV e⁺e⁻ collider a very attractive option. A detailed discussion of the physics programme for a 500 GeV e⁺e⁻ linear collider can be found in the proceedings of the 1991 ICFA meeting held at Saariselska, Finland, and of the Workshop on Physics and Experiments with Linear e⁺e⁻ Colliders held at Waikoloa, Hawaii in April 1993.

Hence, the current goal of R&D on linear colliders is to understand the best and most economical way to build a machine in the 0.5 TeV range, which could later be upgraded to a higher energy, *i.e.* about 1 to 2 TeV. It is generally believed that, today, the latter machine would represent too great a technological step and would have to be built in a step-wise manner. The potential for physics of an e⁺e⁻ linear collider, as a function of its energy, is indicated in Figure 22.

The generic layout of such a machine is given in Figure 23. It comprises electron/positron sources followed by damping rings, compressor areas and pre-accelerators

Figure 22. **e⁺e⁻ energy frontier at linear colliders as a function of energy and luminosity**

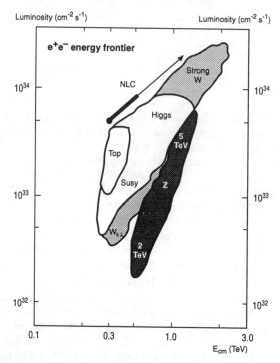

Note: The figure gives a rough estimate of the physics reach for various new particles or phenomena as a function of centre-of-mass energy and luminosity of an EPLC.
Source: D. Burke.

which are necessary to provide the low emittance, *i.e.* beam size and angular spread, of the e⁺e⁻ bunches, which are fed into the main linac.

Linear collider R&D

The SLC experience, the associated R&D on that machine, and other R&D efforts world-wide lead one to believe that most of the technology for generating appropriate beams is at hand now. The major challenges are to optimise and realise the main linac, where high technical requirements will have to be met at reasonable cost, and to design the final focus system, where beams with extremely narrow dimensions have to be generated to achieve the necessary luminosity. Cost and luminosity, in turn, are very sensitive to various machine parameters, such as frequency and acceleration gradient (see

Figure 23. **Generic layout of an e^+e^- linear collider**

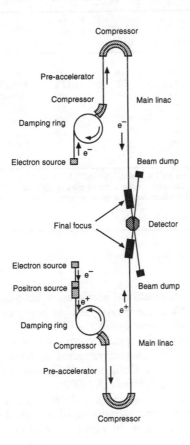

Note: The pre-accelerator section is drawn out of scale, *i.e.* it should appear much shorter than the main linac and the final focus, which would be about 25 km long.
Source: Author.

Table 12). A very significant world-wide effort is being made to address these challenges. At present, there is no agreement as to which technology will be the optimum one, and several technical options are being investigated:

- A far-reaching co-operative venture, involving at present some 20 American, European, and Chinese institutions, is working within the DESY-based TESLA collaboration to demonstrate that a collider based on superconducting niobium structures can be cost-competitive.

Table 12. **Parameters for different approaches for a future e⁺e⁻ linear collider**

	RF frequency (GHz)	Gradient (MV/m)	Rep rate (Hz)	Bunches/ RF pulse	σ_x/σ_y (nm)	P_B (MW)	L (10^{33} cm⁻² s⁻¹)
TESLA	1.3	25	10	800	1 000/64	16.5	6.5
SBLC	3.0	17	50	125	663/29	7.3	3.4
NLC (SLAC)	11.4	38	180	90	300/3	4.2	8.2
JLC (KEK)	2.8	28	50	55	300/3	1.6	4
JLC (KEK)	5.7	40	100	90	260/3	3.6	7
JLC (KEK)	11.4	40	150	90	260/3	3.8	8.1
VLEPP	14.0	96	300	1	2 000/4	2.4	15
CLIC (CERN)	30.0	78	1 700	1	90/8	0.4	2.2

Note: The following nomenclature is often used: 2.8-3 GHz: "S-band"; 5.7 GHz: "C-band"; 11.4 GHz: "X-band". Selected preliminary linear collider parameters for energy in the centre of mass (E_{CM}) = 500 GeV (σ_x and σ_y are the horizontal and vertical beam sizes, P_B is the power of one beam, and L is the luminosity). TESLA uses superconducting RF; all others use room temperature RF.

Source: Author, from information supplied by laboratories.

- A linear collider using conventional acceleration techniques similar to those that have proved their worth at the SLAC SLC is being considered at DESY and Darmstadt (the "S-band" linear collider – SBLC).
- Facilities with a conventional copper structure but working at higher frequency than present machines are being investigated at KEK (Japan Linear Collider – JLC), at SLAC (Next Linear Collider – NLC), and at the Institute for Nuclear Physics at Novosibirsk and Protvino (VLEPP).
- At CERN, research is focusing on a collider (CLIC) working at very high frequency with a power source consisting of a second superconducting linear drive accelerator, which replaces the klystrons used in all the other projects.

When the different approaches are compared, even at the level of rather simple arguments, many partly conflicting parameters must be examined. The situation becomes more difficult if technical problems and the possibility of solving them are taken into account. Accordingly, the number of proposed solutions is quite large (see Table 12). They cover a wide range of parameters, ranging from the superconducting 1.3 GHz TESLA to the 30 GHz CLIC.

The world-wide R&D effort addresses both the question of individual components and that of whole subsystems. Within the framework of the TESLA collaboration, a number of laboratories are addressing the problem of producing high-gradient, low-cost superconducting cavities. At CERN, various aspects of the elements for CLIC are being studied. High-power (50-100 MW) klystrons and advanced RF pulse compression systems are being developed as part of X-band investigations.

Efforts to construct 500 MeV prototype linear colliders for integrated system tests are underway for the TESLA project and for the S-band approach at DESY and for the X-band approach at SLAC. The Accelerator Test Facility (ATF) is now under construction in Japan with international participation. Its main focus is a next-generation damping ring, and it will test the whole pre-acceleration complex from the e⁺e⁻ sources to the main linac. Similar installations are being prepared for the SBLC and TESLA project at DESY and for the NLC at SLAC.

There is a large international R&D effort centred around the Final Focus Test Beam Facility (FFTB) at the end of the linac at SLAC. Its aims are to understand and test the focusing technology required for the very fine beams required in e⁺e⁻ colliders and to develop instrumentation to measure their size. The need for very high luminosity dictates the narrowness of the beam required, which ranges from 100 nm to 3 nm, depending on the technology to be used. The test currently in progress at the FFTB aims to produce 60 nm spot sizes, which already meet some of the requirements noted in of Table 12.

As mentioned in Section 3, R&D on future linear colliders is the object of very active international co-operation. During its last seminar at DESY, in May 1993, ICFA was able to give its support to an interlaboratory collaboration formed to co-ordinate exchange of personnel and information among institutions involved with collider R&D and to review progress towards various collider options. Nearly all major groups involved in the collider work have joined this collaboration, the eventual goal of which is an electron-positron linear collider (EPLC) of very high luminosity in the energy range of 500 GeV, upgradable to 1.0-1.5 TeV. Other laboratories that are making progress in this field have been invited to participate. This is in full accord with the recommendation, often reiterated by ICFA, that for future very large machines, international co-operation should be set up in an early phase and that the world scientific community should agree on a rather concrete proposal before further steps are taken (see also Chapter 6).

Following these recommendations, there is now intense co-operation between the laboratories involved in R&D, which are exploring very different paths. It will take a few more years before the best solutions emerge and their feasibility is demonstrated in the test installations now being built.

5. Summary

In the past two decades colliding beam experiments have been the most fruitful ones in high energy physics. It is therefore not surprising that all plans for future accelerator projects in this field are of this collider (storage ring) type. Progress is obtained either by increasing the luminosity of so-called factories or by increasing the energy.

If attention is solely focused on colliders with a more than 10 GeV centre-of-mass energy, all medium-energy high-luminosity colliders are of the B factory type. Two of these have recently been approved for construction, one at SLAC in the United States and the other at KEK in Japan. Both will be asymmetric colliders, to allow measurement of the separation of the decay vertices of the B mesons, and both aim for luminosities of several times 10^{33} cm^{-2} s^{-1} to supply the B mesons required to study the effect of CP violation in the B$\overline{\text{B}}$ system.

At the energy frontier, research is advancing on both pp and e⁺e⁻ collisions. Consistency arguments within the Standard Model suggest that new phenomena should appear at the energy scale of a few TeV. Since protons are aggregates of quarks and gluons, each carrying only a fraction of the total momentum, the energy necessary in pp collisions would be several tens of TeV. The only project at this energy ready for approval in the near future is the LHC at CERN.[1] It will operate at an unprecedented energy (14 TeV centre of mass) and luminosity (about 10^{34} cm^{-2} s^{-1}). This will put very stringent demands on the technology of the machine, particularly the superconducting magnets, and the detectors. According to present plans, the physics experiments could begin in 2003 provided sufficient additional resources (some 20 per cent of the total cost of SF 2.5 billion) can be found for the project. Otherwise the project would be delayed by about two years.

At a few TeV, e⁺e⁻ collisions should present an equally powerful tool to study the energy frontier. In many respects, they would complement pp collisions. The next generation of e⁺e⁻ colliders, which would exceed the energy of LEP-II (about 190 GeV), must be of the linear collider type. This new technique, pioneered at SLAC, still requires substantial R&D work before a definitive proposal can be formulated. A world-wide collaboration has been set up to optimise and test the required techniques. It will be a few years before the best solutions emerge and their feasibility has been demonstrated. Current efforts concentrate on a machine of about 500 GeV centre-of-mass energy, which would provide an excellent tool to study the top quark and provide information complementary to that provided by the LHC. For reasons mentioned above, the energy should be extendable to 1 to 2 TeV.

---

1. The CERN Council approved the LHC in December 1994.

Chapter 6

Scenarios for Implementing Future Accelerators and Detectors

1. Issues

There are few precedents for collaboration on very complex technological projects on the scale now contemplated. In setting forth scenarios for implementing a collaboration, the first issue is organisational structure. This chapter will describe a number of possibilities that have been tried or have been discussed at international meetings of accelerator physicists. Attention will be given to considerations affecting the choice of organisational structure and to the wide variety of bodies and communities to be consulted so that the solution chosen will receive the widest support. The important practical and political role of existing national laboratories and the balance between bilateral and multilateral discussions will also be brought out. Finally, scenarios will be considered in the context of the actual situation in the field, with comments on the steps that might be taken in order to implement them.

2. Types of organisation considered

The following discussion adopts ICFA's helpful classification of different organisational models, as set forth in a document of May 1993.

"*a)* National or regional facilities:
They should be built and operated by the host country or host region. The planning, project definition and the choice of parameters should be done in international collaboration.

"*b)* 'Larger' facilities which cannot be funded by one country/region:
Here the host country or region should seek contributions for the construction from other countries or institutes. These contributions should preferably be provided in components or subsystems. The facility should again be planned and defined internationally. The operation should be the responsibility of the host country [this is essentially the so-called HERA model].

"*c)* Very large projects needing a collaboration of several countries with comparable share of the total construction and operation cost:

In this model the participating countries should make their contributions again through components or subsystems in a similar way as large collaborations building jointly a major detector facility. A facility under this model would be a common property of the participating countries or laboratories. They would also share the responsibility and the cost for operation. The staff for the operation would also come from the participating countries. To set up this type of collaboration, governmental agreements are expected to be needed.

"*d)* Very large projects in the frame of an international organisation:
This type of collaboration is characterised by common funding for the construction and for operation of large projects. CERN is a prominent example."

It may be useful to elaborate on the description of these four models.

In the past, most projects have used model *a)*, but they have not always been chosen in international collaboration. Still, regional bodies such as the European Committee for Future Accelerators (ECFA) in Europe and the High Energy Physics Advisory Panel (HEPAP) in the United States have usually been consulted, in order to ensure the best possible use of resources. The B factories in Japan and the United States met the expectations of a large scientific community.

Smaller projects have often been duplicated in different regions, in a spirit of competition, even a race to reach some very definite goal; this may not be a bad thing, as it adds to the excitement and to pride in the field. At a certain scale, however, this becomes an unaffordable luxury, and it is that situation that motivated the present study.

Model *b)* has succeeded when the contributions required of outside participants did not constitute too large a fraction of the project's cost. International contributions also serve to validate a project by demonstrating world-wide interest in the research goals. A facility of the size of HERA may be the limit for success with this model, as it becomes more difficult to implement for larger facilities. When international collaboration is discussed in advance, potential participants may have a greater voice in the running of a facility than they do at present under this option.

Model *c)* is used for collaboration on large detectors. Detector collaborations use both the facilities at home institutions and the infrastructure of laboratories at which experiments are performed. So far, this type of organisation has not been used for constructing an accelerator.

3. Variables bearing on the choice of organisational structure

A number of variables affect the choice of organisational structure.

Size

Size is by far the most important variable, since it is much simpler to undertake a project on a national scale when that is possible. Both national users and national authorities, who can more easily justify the expenditure to the public, welcome such a decision. Thus, the driving force behind international collaboration is generally the impossibility for a given nation, or even region, to build a facility beyond a certain size. However, small facilities can sometimes be mounted much more efficiently at large interregional or international centres, so that the decision is taken to implement them there. For example, an accelerator such as ISOLDE at CERN can be used as a source of particles at very low incremental cost; special problems, such as safety, may be more easy to handle at a large installation with a proven record of success; or a large centre may provide a specialised team that would be hard to assemble for a small project.

Complexity of interfaces

Some large projects are modular enough so that a clean interface between large subsystems is possible; they can then be divided up among various collaborators who manage their own subsystems. This is the case, for example, for some aerospace projects, which include a launch vehicle and a spacecraft and thus a relatively easily controlled interface. However, model c) international collaborations are difficult for an integrated project lacking simple interfaces. Accelerators are judged by their success in reliably producing high-quality beams, and this involves a very high degree of interrelation among the components. As a result, experienced accelerator experts are extremely cautious about defining interfaces and consequently wary of model c) organisation. This issue affects the choice of items split off from the main project in the HERA model, and indeed it may impose limits on which parts are allowed to be handled in collaboration. In model d) organisations, where there is undivided responsibility, as at CERN, there is a proven record of success.

Timetable for the project and operating costs

Negotiation procedures may be quite different for one-time contributions of equipment or funds for constructing an accelerator and for projects involving contributions, which may evolve, over a number of years. The latter case requires a long-term commitment, and it is helpful to include flexibility when defining the project. Since the annual operating costs of an accelerator facility are not much below the annual expenditure during construction, operation is likely to be a problem if a large part of the construction costs was shared. This is another reason for considering organisations of models c) and d), where long-term commitment is clearly established.

Pre-existing facilities

It may be that a new facility, though very large, is nonetheless a fractional increment to an existing facility. The LHC is a clear example, but other new projects may fall into this category. In this case, the structure of the existing laboratory must be an important consideration when choosing an organisation for a new project. This does not mean that a new organisation cannot be created, but it means that it must have an appropriate relation to the existing one.

Site

The nature of the host country, or region, must be taken into account. The case of a small country that cannot bear a major fraction of the cost has to be considered differently from that of a country that pays a substantial portion of the total. Taking the benefit of spinoff effects into account, it is natural to expect the host country or countries to contribute considerably more than others.

There are many advantages to placing a new facility at an existing laboratory. Both prior experience and the existing infrastructure can be extremely valuable for the new project. In some cases, space limitations, power and cooling demands, or geological features could make such a choice impractical.

Site selection is one of the most difficult problems for international collaboration on future accelerators, because there are likely to be more "losers" than "winners". The "multiple megaproject approach" could considerably alleviate this problem if several megaprojects were dealt with at the same time, and thus be considered to be hosted by different countries.

In this approach, megaprojects may involve a range of scientific or technical activities that goes beyond high energy physics, thereby offering more possibilities for a better distribution of opportunities among different regions. Combining megaprojects in quite different disciplines, however, would create additional difficulties. It would require discussions among members of different disciplines, and if projects from both basic and applied sciences were included, it would be difficult to evaluate and compare them using common standards. Furthermore, at any given time, individual projects are at generally different levels of readiness. Selecting a site before the design is completed would mean basing the choice mainly on political factors. This might prevent obtaining the best match between site and technology and might also have a detrimental effect on the formation of a broad collaboration.

Different laboratories often develop different techniques for a new accelerator. Site selection might depend on the success of a particular laboratory in developing the technology for the future facility.

Political goals and political barriers

In some cases, the contribution to a project is made easier by political considerations in the countries involved, either because they have particularly warm political relations or because they wish to couple the issue of collaboration to other matters at a particular time. This fits well with model *b)*. There are certainly cases of scientific collaboration among nations whose political relations are not particularly warm as well. This might be expected to be a barrier, and may indeed be a barrier to some types of collaboration. However, scientific collaboration can sometimes be an element in achieving a broader political goal. The ancient paradigm for this mechanism is the Olympic games, which brought together people from hostile States. In a very different context, scientific collaboration has been shown to achieve the same goal. If this motivation is important, the organisation most able to advance this goal is likely to be model *d)*.

4. Entities with a voice in the choice of a plan

Many constituencies will be involved in the discussions leading to the formation of the plan to implement an international project, or even the type of co-ordination needed in projects using model *a)*. First, there are the scientists who are to use the new facility, including the theoretical physicists who benefit from the results and usually support the new projects quite actively. They are responsible for demonstrating that the science is valid, feasible, and worth the cost. This group will naturally be particularly concerned to have the project succeed. The actual implementation will not be in their hands, but if the aims of the project drift away from their goals, their support will soon evaporate.

The accelerator scientists and engineers are a most important community, as the success of the project depends on them and as none of the new projects presently planned is easy to realise. They must work in an organisation that respects them and their views, and it is utterly wrong to assume that once the project is well-defined, it can be taken out of the hands of the experts and turned over to industry. They will insist on a clear and accountable management structure.

Both of these constituencies are represented by national professional organisations, which have no official standing but are influential by virtue of the stature of their members. They usually set up panels, workshops, and the like to discuss the project.

There are advisory panels with a mandate from the physics community, such as ECFA in Europe, or a government mandate, such as HEPAP in the United States. Their specific task is to give advice on future projects in the field. Their advice is particularly sought on scientific and technical matters, but they will no doubt comment on organisational matters as well.

Regional and national learned societies such as the European Physical Society (EPS) and the American Physical Society (APS) provide a forum for discussion of the high energy physics programme in the framework of physics as a whole.

The national funding agencies are a primary focus of the negotiations, since the ultimate funding may come from their budgets. Even if it does not, but goes through a Ministry of Foreign Affairs, for example, it is laid to their charge. In many countries more than one funding agency is involved.

A Ministry of Science, or the equivalent, must be expected to play a major role in a large new international scientific endeavour. Since the political goals and barriers mentioned above may be an issue, other ministries are also likely to be involved.

Parliaments are inevitably involved in the discussion of the large expenditures required for megaprojects. They will also usually be interested in the political aspects of the problem.

In high energy physics, international bodies, both regional and global, already exist. They have played an important role in forming high energy physics programmes and will continue to do so. Consultations with them will be an essential part of the process of forming future programmes. In 1976, the International Committee for Future Accelerators (ICFA) was created by IUPAP primarily as a forum for discussing the world high energy physics programme. In recent times, discussions at this forum have intensified and led to a very clear formulation of the future steps that could be taken in a world-wide programme. In light of the complexity of the issues involved, it is important to initiate discussions between government agencies in different countries and the world-wide scientific community. The recently constituted OECD Megascience Forum can play a very important role in organising such contacts and facilitating an exchange of views.

In fact, all of the parties mentioned above are active at the present time, given the urgency resulting from the cancellation of the SSC project and the culmination of the decision process on the LHC, with the concomitant need to consider the status of other parts of the high energy physics enterprise.

5. Multi-disciplinary laboratories as sites for large accelerators

Increasing specialisation in science is probably unavoidable. Nevertheless, it is sometimes possible to combine the interests of different scientific communities in a single project. For example, solid state physicists and biologists use synchrotron radiation from electron-positron colliders. Multi-disciplinary use of future facilities should be strongly supported. The electron-positron linear collider (EPLC) can perhaps be used not only for physics but also for biology.

Many large accelerators are at laboratories set up solely to build that facility, essentially for one field of research. These are called "single-purpose laboratories", even if other applications of the facility almost always develop in time. Other laboratories have instead been founded explicitly as multi-disciplinary centres covering a broad spectrum of pure and applied research, with the idea that the grouping of scholars and research facilities will lead to a whole that is greater than the sum of its parts.

If new facilities are based on existing laboratories – a preferred strategy – a single-purpose laboratory might move towards becoming multi-disciplinary. In that case, support would come from more than one research community. One activity might be high energy physics, another might be nuclear physics. In the past, these two communities were served by different facilities, but as they start to use the same ones, the advantages of shared facilities become clear. One important advantage is that the closing down of any one facility, or even one line of research, does not threaten the existence of the laboratory. This leads to a flexibility of operation that helps solve a big problem in the field – how to *close* a facility. This is a powerful argument for siting a large accelerator at such a laboratory.

6. Basic elements of international collaboration

Agreements between research teams and laboratories should continue to form the basis of international collaboration in high energy physics. HERA, which was built mainly on the basis of inter-laboratory agreements (supported by the governments of participating countries), offers an excellent example of such collaboration. However, the role of government agreements increases rapidly with the size of the facility, and they will be increasingly important for future projects. It may be necessary to have not only intergovernmental but even interregional agreements. If priorities are first discussed within regions (*e.g.* Europe, America, and Asia) it may be easier to reach a global agreement.

7. Future role of ICFA and ECFA

International committees such as ICFA and ECFA play an important role in the organisation of scientific, technical, and sociological discussions of future accelerators (see Chapter 4, Section 8). Such discussions will be even more important in the future, because of the larger scale of facilities and the likelihood that they will be constructed in a more international mode.

It may be possible to transform ICFA and ECFA into bodies for defining priorities for future facilities, choice of technologies, and site selection at global and regional levels. Their proposals could be then considered by governments.

This would probably require ICFA to have a broader representation of user communities. The chairman of ECFA already represents the European high energy physics community in ICFA. If analogous user representatives from other regions (America, Japan, Russia, etc.) were added, it would have a balanced representation of users and directors of major laboratories and even greater authority than it has at present. It would also be desirable to develop clearer procedures for nominating representatives of certain regions to ICFA.

A transparent and well-defined decision-making mechanism should be established long in advance of the time of actual decisions.

8. Megaprojects and smaller projects

Projects in high energy physics cannot be exclusively megaprojects. Because it is impossible to predict which direction in basic science will bring the biggest surprises, as many approaches as possible should be followed. Not all outstanding questions can be answered by megaprojects, and some are better addressed by smaller facilities. A phi meson factory or a tau charm factory are good examples. Other examples might be a meson factory or non-accelerator experiments. These projects usually also involve one type of international collaboration or another and should receive adequate support from the scientific community and governments.

9. Scenarios for B factories

The issues discussed here are the organisation of the proposed B factories, the reasons for their number, and their relation with other B physics facilities.

Organisational structure of the planned facilities

The B factories at SLAC and KEK are being built as national facilities using model *a)*. Yet it is likely that other countries, such as Canada for the SLAC project and the countries of south-east Asia for the KEK project, will be attracted to them.

At the present time, discussions are taking place between SLAC and Russian and Chinese laboratories that might construct some of the PEP-II accelerator components. It appears very likely that some mutually beneficial arrangement will be reached and that institutes from these countries will collaborate on the PEP-II construction. China and Russia have broad experience in the construction and exploitation of medium-energy electron-positron colliders and could be potential partners in both cases. They can construct and deliver parts of the accelerators and detectors for prices considerably below world prices. The price differential could be considered as their contribution to the projects (*i.e.* a "partial compensation" approach; see also Chapter 4, Section 6). Discussions on Russian participation in the KEK B factory are taking place as well.

The advantages of international collaboration have been discussed in Chapter 4. These laboratories, both of which hope to play an important role in the EPLC, could use, in the future, the experience gained in international collaboration on the construction of a B factory. However, unnecessary delays in approved projects, due to the involvement of other countries, should be avoided. Additional contributions can be used to speed up projects or to remove compromises made in the design because of limited funding.

The number of B factories

It is appropriate to ask whether the construction of two B factories constitutes an unnecessary duplication, but there are sound scientific arguments in support of having two factories.

For technical reasons, it is not possible for either B factory to have more than one detector. Arguments in favour of having at least two experimental set-ups and groups of physicists include:

- complementary approaches and techniques in the challenging study of a new physics domain, especially in these accelerators which require very high luminosity;
- confirmation of results obtained using different set-ups;
- scientific competition between teams, an essential part of science.

The effectiveness of such an approach was demonstrated by the simultaneous operation of the CESR and DORIS electron-positron colliders over more than a decade. It resulted in rapid development in different fields, especially of B physics and tau lepton physics. The observation of unexpectedly large mixing between neutral B mesons was the basis for the B factory projects.

Relation with other B physics facilities

As already discussed in Chapters 1 and 5, B physics is a very important direction in high energy physics, as it provides information on fundamental parameters of the Standard Model and can be sensitive to effects beyond this model. Several scenarios for developing this field have been discussed by the high energy physics community. Technical studies of the possibility of building B factories were carried out in many laboratories (PSI, CERN, DESY, Novosibirsk, Cornell, SLAC, and KEK). The idea that such a facility should be built somewhere was strongly supported by the consulting bodies of the scientific community, and two projects obtained approval. Other laboratories concentrated on different projects, including complementary approaches to B physics studies.

Studies of B physics at existing accelerators (CESR, HERA, LEP, SLC, Tevatron) will continue during the construction of the B factories and in some cases during their operation. These studies are expected to provide important complementary information in this very rich field. In particular, experiments at the Tevatron and HERA will be an important step towards the investigation of B physics at LHC.

It may be useful to recapitulate here the relations between the physics capabilities of these many different facilities. CESR offers a clean approach to B physics in electron-positron annihilations in a facility with a programme of steadily increasing luminosity, but it lacks the capability for the asymmetric collisions necessary for measurements of B decay times required for CP violation studies. It is this feature that is unique to the new B factories under construction. The other machines that use electron-positron collisions, LEP and SLC, are unlikely to achieve the high rates required for a definitive study of CP violation. The other machines involve hadrons, and this leads to a much more difficult environment, but they can offer very high rates indeed and, in the end, are likely to give essential information available from no other source.

10. The Large Hadron Collider

The LHC was planned as a further large facility in the framework of the CERN accelerator complex. It will make use of the existing LEP tunnel and of most of the CERN infrastructure. Thus, it fits quite naturally into ICFA's model *d)*. However, supplementary participation by non-CERN member countries has been considered from the very beginning of the project. After the demise of the SSC, this international and interregional participation has become an even more important part of the overall planning.

It is certain that the LHC project will attract a large number of scientists from all over the world. Several hundred of those scientists are already actively taking part in the planning and design of the large detectors at the LHC, and this number is expected to grow. Although such a facility should, in principle, be open to all scientists and research teams, it is generally agreed that the massive use by non-member country scientists envisaged at the LHC must involve some contribution by these countries to the construction and perhaps the running of the machine. The mechanisms for these contributions should certainly be flexible and take into account both the needs of the LHC and the possibilities of the countries concerned.

Contributions should be negotiated in a spirit of partnership, with the understanding that non-member country contributions are important to the success of the project. If formal requirements for contributions to the constructions costs by non-member countries are introduced as a condition for participation in the research, this may trigger demands for reciprocal contributions by European scientists who use non-European facilities. These demands would not necessarily be limited to particle physics and could have disastrous consequences for the global exchange of scientists.

It is appropriate to recognise major contributions to the machine by non-member countries by involving the latter both in the direction of the project and in official CERN bodies (see below).

In all this, experience gained from the HERA model should be taken into account. Thus, the LHC project could be considered as ICFA's model *d)* with some elements of model *b)*. It is possible that this kind of international collaboration could serve as a model for future global international research.

On the proposed timetable, the LHC project (together with the rest of the CERN programme) requires some SF 500 million in addition to the presently assumed level of member country contributions of approximately SF 975 million per year. These extra funds could come from:

– contributions from non-member countries;
– special additional contributions from some member countries, such as the CERN host countries.

If these funds are not found, additional delays and severe sacrifices in the scope of the project are unavoidable.

Contributions by non-member countries to the LHC project

The question of non-member countries' contributions to the LHC project is an extremely difficult and delicate one. It is vital to solve this problem if the project is to be completed according to the proposed timetable. Several scenarios for developing appropriate mechanisms are possible:

i) Co-operation agreements:
This approach has already been approved by the CERN Council. Co-operation agreements have been concluded with several non-member countries in Europe, in Asia (China, India), in Latin America, and with Australia. These agreements provide the framework and legal basis for co-operation: additional protocols specify the details of their implementation. A new protocol on Russian participation in the LHC project is under discussion. Possible co-operation agreements between CERN and several other countries are being discussed.

It is appropriate for countries that have signed co-operation agreements with CERN to become observers at CERN. Observer status simplifies the exchange of information and helps to build mutual understanding. It does not, however, provide a mechanism for a country's participation in CERN activities and does not require such participation.

The co-operation agreement has several advantages:
– It is already an established practice.
– It is flexible enough to accommodate different sorts of co-operation (contributions in cash, in kind, using "partial compensation" approach, etc.) and differences between countries.
– It can delegate the execution of protocols to individual institutions and thus to the physicists who are most interested in the success of the project.
– It does not require additional administrative structures.
– Signed at the government level, co-operation agreements provide some guarantees that obligations will be met.

There are, however, some limitations:
– The scale of existing agreements is an order of magnitude smaller than what is required for the LHC project. So far, they have been fulfilled successfully, but it is not clear whether this mechanism is firm enough for large-scale commitments.
– It does not allow non-member countries to influence the decision-making process, which countries making a large contribution would naturally wish to do.

ii) Associate status:
For larger contributions, it may therefore be appropriate to consider an "associate status" scheme. Its purpose would be to allow a non-European country to develop an institutional partnership with CERN and to be closely involved in specific CERN activities. The rules governing associates should be flexible, in order to recognise the differences among countries in terms of their use of CERN, their economic strength, and their reciprocal facilities. They should respect, as much as possible, the spirit of international scientific co-operation and the importance of a global exchange of scientists.

The benefits of associate status should be sufficiently attractive for a country to make a contribution to CERN. For example, assuming a sufficiently large cash contribution, they could include the rights for the associate country's firms to take part in CERN tendering procedures, to propose candidates for fellow and student posts, and to participate in discussions of CERN programmes.

Associate status would allow for greater integration of a non-member countries in the LHC project than was previously possible and should help ensure that contributions to the project are not vulnerable to economic and political changes. The concept is still being developed at CERN and seems the appropriate way forward.

iii) Membership in CERN:
Membership in CERN is a natural perspective for a European country. At present, however, it is hard to identify any European non-member countries that could make a considerable impact on the LHC programme by joining CERN.

The possibility of non-European countries becoming members of CERN is unlikely in the near future because:

- It would require a change in the present interpretation of the CERN Convention, perhaps a revision of the Convention.
- Under present rules, both the United States and Japan would have to contribute 25 per cent of CERN's income. This would require a complete rethinking of their domestic programmes, and it would reduce the role of European countries in CERN to a level probably unacceptable to at least some of them.
- Membership in CERN has long-term consequences that go much beyond the LHC. Presumably, both scientific communities and governments would be more hesitant to make general long-term commitments than they would be to make commitments only to the LHC.
- Even with good will on all sides, such a drastic change would require long and difficult negotiations that would be incompatible with the LHC timetable. Therefore, keeping in mind that CERN may become a world laboratory at some future date, this possibility will not be discussed further in the context of the LHC project.

Participation of the world scientific community in the LHC project

Interest in participation in the LHC project has already been expressed by hundreds of physicists from non-member countries. Many are already heavily involved in work on the project. Taking this interest into account, many countries, including Canada, China, India, Japan, Russia, and the United States, are now considering the possibility of contributing to the LHC project. In some cases, positive decisions are expected in the near future.

As regards the US high energy physics community, the investigation of the super-high energy frontier, using proton-proton collisions, has been its primary goal for over a decade now. In the wake of the cancellation of the SSC, the LHC provides the only way to study this physics in the foreseeable future. Therefore, it is natural to anticipate the involvement of many US scientists in the LHC project. Interest in participating in the LHC has been already expressed by many US experimentalists and accelerator physicists.

Participation in the LHC project would allow the United States to use very efficiently the experience gained during the preparation of the SSC. Presumably, a large share of the US contribution to the construction of the LHC could be made in kind, using technologies developed for the SSC.

The major obstacle to US participation in the LHC is the drastic reduction in total funding for high energy physics in the United States. Not only was the SSC cancelled, but severe cuts were made in the US high energy physics operating budget, corresponding to about a 20 per cent loss in purchasing power over the last three years (FY 1993-95).

Another complication is the uncertainty of US commitments to long-term projects, as project funding can be terminated every year. The SSC is the most significant example. A high-level international agreement between CERN and the US government could help alleviate this problem, by setting up a context in which respecting such a commitment has high priority.

As regards Japan, which has developed a high level of technology and know-how at KEK, it could be a possible contributor to LHC. At this moment, Japan's position on the LHC is not clear, but further internal discussion is expected to take place, depending on CERN's decision to build LHC.

Timetable for decisions on the LHC

Taking into account the objective difficulties in reaching early agreement on non-member countries' contributions, CERN management follows the so-called "planning-for-success" approach. It assumes that the LHC programme will be defined and approved, on the basis of whatever level of contributions member countries are prepared to provide and whatever extra income can be foreseen. The expected additional contributions will make it possible to carry out the full LHC programme. By a variety of measures, the most important being to stretch the LHC construction time, it is possible to formulate a workable scenario for the improbable case that no contributions from non-member countries are forthcoming. Scenarios with different funding levels will only start to diverge in 1997. This means that there will be time for negotiations on additional contributions once the LHC has been approved.

Approval of the LHC in 1994 would make it possible:

– to maintain the momentum of the project;
– to sustain the efforts of the many hundreds of physicists and engineers who are working hard on developing new techniques for the LHC;
– to plan the use of resources in the coming decade optimally;
– to maintain the interest of industry, which is carrying out development work in anticipation of the approval of the project;
– to simplify negotiations on contributions from non-member countries.

Postponing approval of the LHC project until contributions from non-member countries are fixed would make the corresponding negotiations much more difficult (if not impossible). It would also eliminate other advantages of early approval noted above, and would in fact jeopardise the entire LHC project.

Composition of CERN advisory committees

Non-member country scientists are increasingly represented on these committees. This trend should help enlarge the experience being tapped and should be encouraged. It will help the LHC profit from knowledge gained in the construction of the SSC and its experiments and should also help the non-member countries find the best way to participate in the LHC.

11. An electron-positron linear collider as the next large facility

The great push to create a facility for high luminosity hadron collisions at the energy frontier has been accompanied by the presumption that the obvious complementary machine, an electron-positron linear collider (EPLC) with an initial energy of around 500 GeV, should also be constructed (see Chapter 5).

Several issues will influence the scenarios for implementing an EPLC, including the development of R&D on different technologies, the choice of technology, the extendability of the EPLC energy, and the form of organisation of the project.

World-wide convergence on a common project

As already mentioned in Chapters 4 and 5, R&D on future EPLCs is being performed, to a large extent, in international collaboration. ICFA recommends establishing international collaboration at the R&D stage for large accelerators; it also recommends that the scientific community should formulate a relatively well-defined proposal, including realistic cost estimates and site criteria, before formally approaching governments. International collaboration on R&D should simplify collaboration during the construction phase using one, or a combination, of the models proposed by ICFA, as discussed below.

It would be extremely desirable for scientists to reach a consensus on co-operation mechanisms. Different ways to reach this goal can be envisaged. The decision-making committee could be formed by interested parties close to the end of the R&D phase. Alternatively, the bodies created to steer international collaboration on R&D could gradually be transformed into decision-making bodies at the level of the scientific community. However developed, the committee should define parameters and develop procedures for choosing the technology, as well as propose site selection criteria. Many obvious needs will determine the choice of technology, among them feasibility, cost-effectiveness, and reliability. One very important criterion for technology and site selection is a capability for subsequent upgrade to a higher energy. The major problem is to find a reasonable balance between different requirements. Here, the large amount of experience with procedures for choosing technology for constructing large detectors can be of use.

When the technology is chosen and the site criteria are established, governments can be formally approached with a request for the creation of a new international organisation or collaboration structure. Obviously, informal consultations with governments during the entire preparatory stage are essential to success. Recommendations from the scientific community would simplify the task of governments when a new organisation is to be created. On the other hand, those recommendations can be a severe constraint: it is natural to place the new organisation at the laboratory that developed the technology selected.

This scenario assumes that differences among technologies are great enough to make it possible to make a clear choice among them. That may not be the case, and the different R&D programmes may not be equally ready when the choice is to be made. This suggests that, if an international organisation is to be created, it should be created at an earlier stage: in this case, the site would be selected mainly on political (economic) criteria. In this scenario, a gradual increase in the organisation budget could simplify its creation.

A detector at the EPLC will be highly integrated with the accelerator, and joint collaboration on the detector and the accelerator would be a possibility. This would unite the communities of experimentalists and accelerator physicists in their struggle for approval of the project. It would also diversify possibilities for contributions to the project from different countries.

The advantages and disadvantages of the different types of organisation are set out below, starting with model *d)*, which represents the most advanced level of collaboration.

International organisation, model d)

Among the main advantages of this model for the EPLC can be mentioned:
- existence of a very successful example, as many large facilities have been selected, designed, constructed, and successfully operated by CERN;
- well-defined long-term commitments by the member countries of this international organisation;
- effective procedures for steering the project;
- clear mechanisms for operating the facility;
- possibility to use the organisation for extensions of the project and new projects.

There are also limitations and problems in such an approach:
- The need for one more international high energy physics laboratory is not clear. Presumably, governments would be more reluctant to make long-term commitments than to agree on funding a project within a limited time period.
- CERN's success as a regional organisation cannot be automatically extrapolated to the world-wide organisation that is probably required to build the EPLC. The CERN member countries have serious problems in reaching consensus, and some have difficulties in fulfilling their financial commitments. Such problems could be even more serious in a larger organisation with more diverse interests.
- Creating a new organisation when defining the project could be problematic, but creating it once the project is defined would probably take a long time and lead to a loss of the momentum gained during the R&D stage. Creating the organisation before the project is defined appears even more problematic.
- It will be difficult to find a balance between the influence of small countries in the organisation and their contributions.
- Creating a new organisation raises the problem of its future after the project is completed.

Probably the major problem for creating a new international organisation for constructing the EPLC is the need to discuss simultaneously funding mechanisms, site, and choice of technology. The elaborate mechanisms for determining scientific policy which exist at CERN and constitute one of the advantages of an international organisation would not yet be available. As a result, political reasons might dominate scientific ones. Simultaneous discussion of several megaprojects could alleviate the problem of site selection, but might obscure issues such as choice of technology. In any case, it is desirable to establish the decision-making mechanism and procedures early.

The international organisation could be created without the adhesion of every potential member country active in this field of research, once enough countries have joined to pass the critical threshold. Other interested countries could then join the project on the basis of membership or co-operation agreements. However, it is important that all countries with important activities in the field of EPLC join the organisation. Otherwise, it would be difficult to co-ordinate the development of the field.

CERN is the only *international* organisation in high energy physics at the present moment. For more than a decade, however, CERN will be occupied with the construction and commissioning of the LHC. A new international organisation for EPLC could be created on the basis of an existing *national* laboratory. This would reduce expenses for developing the infrastructure and speed up the creation of the organisation.

International collaboration, model c)

This model has not been used so far for accelerator construction. However, there is enormous experience with this kind of organisation in constructing large detectors, including the initial stages of work on the very large detectors for the SSC and the LHC. Detector collaboration has shown that it is possible to:
- make difficult choices for technology without destroying the collaboration;
- develop the organisational structure of the collaboration as work proceeds on defining the project;
- work actively to attract funds for building the detector from different countries;
- design experiments under somewhat uncertain funding conditions;
- distribute work among the collaborating institutions and organise the necessary cash flow;
- operate the detector efficiently for many years.

All these capabilities match nicely the requirements for constructing the EPLC. However, there are important differences between detector and accelerator collaborations:
- The scale of detector projects is much smaller than that of the EPLC.
- Collaboration for detectors usually forms once the general concept has been defined. For the EPLC, the general concept has yet to be defined.
- A large share of the equipment for a detector is usually produced by collaborating institutions. A large part of the equipment for the EPLC is expected to come from industry.
- The sums required during the operation of experiments are usually much smaller than those required annually during construction of the detector. The difference is less pronounced in the case of accelerators.
- Site selection has never been an issue for detector collaboration.

- Detector collaborations rely heavily on the host laboratory's infrastructure. More-over, a significant fraction of the detector cost is usually covered by the host laboratory. For example, CERN contributed 25 per cent of the LEP detectors. A similar fraction is expected for the LHC detectors.

The last difference is not intrinsic. It is possible to imagine a scenario in which one country offers, as a contribution to the collaboration, support in the form of the infrastruc-ture of an existing laboratory. The laboratory would not necessarily be absorbed by the collaboration. This approach would be especially attractive for a country that needs to reduce the number of its laboratories. The country could also offer a large contribution to the running of the EPLC, for example by providing the staff and thus alleviating the problem of staff recruitment in other countries (as well as the domestic unemployment problem).

The main advantages of model c) for EPLC are:
- It offers greater flexibility than the international organisation, and it is possible to optimise the structure for one project.
- Shorter-term commitments would presumably be more acceptable to governments.
- There would be well-established procedures for technology choices.
- Scientists would be broadly involved in the decision-making process.
- Administrative staff would be reduced (by putting more load on the administra-tive staff of the existing laboratories).
- It would offer well-established procedures for finding a balance between contribu-tions (including intellectual ones) to the collaboration and influence in the collab-oration. The problem of small countries would not apply. The unit in a collabora-tion is a research team, not a country.

Among the disadvantages can be identified:
- Experience at the required scale does not exist.
- Site selection remains an open problem, although the selection criteria can be formulated.
- Running the facility could be a problem. Even at the detector level, collection of funds for operational expenses is not trivial.
- It would be difficult to define how to divide the construction into parts that can be contributed in kind by the collaborating institutions (countries).

However, a large commitment by one country and involvement of an existing high energy physics laboratory would be very valuable to the success of model c).

National facilities, models a) and b)

Models b) and especially a) have the simplest organisation. If one country could afford a large expenditure on the EPLC, many issues, such as site selection, technology choice, running the facility, etc., would be greatly simplified, possibly at the expense of foregoing the best choice. Under some circumstances, it is easier to convince a government to approve a national project. Thus, these models could, in some cases, form the organisational basis of the EPLC. However, the planning, project definition, and choice of parameters should be accomplished in international collaboration. In any case, collaboration on the detector should be international.

Model *a)* presents some disadvantages; they are less important in model *b)*:

- The main problem is the difficulty of co-ordinating high energy physics research at a world-wide level. Several countries could start their own EPLC projects. The resulting competition could lead to the abandonment of other, perhaps more interesting projects.
- It is more difficult to choose technology on a rational basis in a national project, and technology developed in the country is most likely to be selected.
- It is more difficult to attract experts from other countries to a national project.
- A national project may be politically and financially less stable than an international project.

Summary on EPLC implementation

The four organisational models for EPLC each have advantages and problems, the weights of which depend on factors beyond our present knowledge. Any one or a combination could work well in the right circumstances, but it is not possible to designate one as best at the present time. Experience gained in the practical realisation of the LHC project will definitely influence the choice of a scenario for EPLC.

12. Scenarios beyond the LHC and the first EPLC

This discussion looks so far to the future that any remarks must be considered highly speculative, but it is an aim of this report to set up a framework for planning projects before they become frozen in institutional frameworks, which tends to happen very early in their development.

The goal of a 2 TeV EPLC has often been discussed as the next logical step in the development of the field. It is possible that the technology of such a machine will be different from that for a 500 GeV machine. The results of the ongoing R&D effort will help to settle this issue. Another possibility would be a hadron collider of energy considerably higher than that of the LHC. Vigorous R&D should be pursued on both of these options as well as on new accelerator technologies in order to open up new scientific perspectives for future generations.

13. Summary

Effective mechanisms have already been developed for comparing the scientific merits and the cost-effectiveness of different projects in high energy physics. They include extensive scientific discussions and consultations among a wide variety of interested parties: communities of theorists and experimentalists, accelerator physicists and engineers, national and international advisory panels, physical societies, international bodies, funding agencies, and parliaments.

These mechanisms, coupled with some historical developments, have led to a proposal for a balanced and well-founded future programme in high energy physics. It includes upgrades of existing facilities and the construction of new accelerators. Projects for several new facilities are now at different stages.

Two B factories have been approved, and the LHC is on the verge of approval.[1]

Very extensive R&D work on an EPLC is underway, and a proton-proton collider, with an energy considerably higher than that of the LHC, is being discussed.

Organisational structures of planned facilities fit naturally into the classification proposed by ICFA:

– model a): national project;
– model b): national project with contributions from other countries (HERA model);
– model c): international collaboration;
– model d): international organisation.

B factories are being built as national projects [model a)], although the definition of the projects and the choice of parameters were based on broad international discussions. Active negotiations are taking place on possible Chinese and Russian participation in the construction of these facilities. It is still possible that some elements of the HERA model will be incorporated into the organisation of these projects.

It is planned to build the LHC as a new large facility in the framework of CERN. Contributions from non-member countries would allow the project to proceed more rapidly without compromising any of its scientific goals. Several countries, including Canada, China, India, Japan, Russia, and the United States, are now considering their possible contributions. The LHC fits into ICFA's model d) but incorporates some elements of model b).

Participation of non-member countries in constructing the LHC can be organised using the existing practice of co-operation agreements. Closer institutional relations would be appropriate in the case of large contributions; the introduction of an associate status would allow a country to participate in CERN's discussions of scientific policy, to have representation in CERN official bodies, and perhaps to obtain additional benefits. The possibility of non-European countries becoming CERN members seems too remote to be discussed in the context of the LHC.

Approval of the LHC in 1994 is extremely important for maintaining momentum and for simplifying negotiations on contributions from non-member countries.

Vigorous research and development work on the EPLC is being done now to a large extent in the framework of international collaboration. An "Interlaboratory Memorandum

---

1. The CERN Council approved the LHC in December 1994.

of Understanding on Research and Development towards TeV Linear e^+e^- Colliders" has been signed by essentially all the major participating laboratories and universities, and it is supported by ICFA. This complies with the ICFA recommendations that international collaboration should be established at the R&D stage for large accelerators and that the scientific community should formulate a relatively well-defined proposal, including realistic cost estimates and site criteria, before formally approaching governments.

The organisational structure for the EPLC is still a completely open question. All four ICFA models or a combination might be suitable, depending on the circumstances. Their advantages and limitations have been discussed above. There are many reasons to believe that the EPLC will be constructed through international collaboration: the choice of technology, the choice of energy, and particularly the choice of site are the main issues that the particle physics community must face towards the end of this decade.

Site and technology selection are serious problems for reaching consensus on a common project. The "multiple megaproject approach" (see Section 3) could alleviate the problem of site selection through simultaneous consideration of several projects and their balanced distribution in the world. It could be also effective for collecting sufficient funds for future projects and preventing unnecessary duplication. On the other hand, it might aggravate some problems, such as the choice of technology, and create new ones, such as the need for complex discussions among different disciplines. The formation of a broad international collaboration would become more difficult, especially if scientific and technological megaprojects are intermixed. In sum, it will be hard to find an ideal universal strategy, and the best solution should be sought on a case-by-case basis.

The organisational structure of future facilities should maintain the principle that large accelerators should be open to scientists and research teams from all countries, as stated in the ICFA guidelines.

Annex 1

Quantitative Information on Particle Physics Research in OECD Countries

These tables are based on data kindly provided by the National Delegations
of the Megascience Forum, and by the participants in the Expert Meeting on Particle Physics.

Table A.1. **Organisation of particle physics research in OECD countries and in Russia**

	Ministry in charge of particle physics	Councils or institutions	Nature of laboratories	CERN membership
Australia	– Department of Industry, Science and Technology	– Australian Nuclear Science and Technology Organisation (ANSTO)	5 universities	No
Austria	– Federal Ministry for Science and Research	– Foundation for the Promotion of Science and Research – Institute for High Energy Physics of the Austrian Academy of Science	5 universities	Yes
Belgium	– Ministry of Economic Affairs	– Inter-university Institute of Nuclear Sciences (IISN)	5 universities	Yes
Canada		– Natural Sciences and Engineering Research Council – National Research Council	4 universities	No
Denmark	– Ministry of Research	– Accelerator Committee of the Danish Natural Sciences Research Council	3 universities	Yes
Finland	– Ministry of Education	– National Research Institute for High Energy Physics in Finland (SEFT)	8 universities	Yes
France	– Ministry of Higher Education and Research – Ministry of Foreign Affairs (for participation in CERN)	– National Institute for Nuclear and Particle Physics (IN2P3/CNRS)[1] – Atomic Energy Commission (CEA)	11 10 IN2P3/CNRS + universities 1 CEA	Yes
Germany	– Federal Ministry for Research and Technology (BMFT) – Science ministries of the 16 federal states	– Max Planck Society (MPG) – German Research Foundation (DFG)	51 29 university labs 18 theoretical groups 3 MPG institutes 1 DESY	Yes
Greece	n.a.	n.a.		Yes
Ireland	n.a.	n.a.		No

Table A.1. **Organisation of particle physics research in OECD countries and in Russia** *(cont'd)*

	Ministry in charge of particle physics	Councils or institutions	Nature of laboratories	CERN membership
Italy	– Ministry of University and Scientific and Technological Research (MURST)	– National Institute for Nuclear Physics (INFN)[1]	20 18 universities Frascati National Laboratory (LNF) Gran Sasso National Laboratory (LNGS)	Yes
Japan	– Ministry of Education, Science and Culture		KEK Institute for Nuclear Study (INS, Tokyo University) International Center for Elementary Particle Physics (ICEPP, Tokyo University) Number of universities unavailable	No
Mexico		– National Council for Science and Technology (CONACYT)	9 universities	No
Netherlands	– Ministry of Education and Sciences	– Agency for Physics Research in the Netherlands (FOM)	7 6 universities NIKHEF Institute	Yes
Norway	– Royal Ministry of Education, Research and Church Affairs	– Research Council of Norway	3 universities	Yes
Portugal	n.a.	n.a.	n.a.	Yes
Spain	– Interministerial Commission for Science and Technology – Ministry of Education and Science – Ministry of Industry		11 8 universities 1 Ministry of Industry 2 "mixed" – university centres and other agencies	Yes

Table A.1. **Organisation of particle physics research in OECD countries and in Russia** (*cont'd*)

	Ministry in charge of particle physics	Councils or institutions	Nature of laboratories	CERN membership
Sweden	n.a.	– Swedish Natural Sciences Research Council (NFR) – Swedish Council for Planning and Co-ordination of Research (FRN) – Wallenberg Foundation – Swedish Research Council for Engineering Sciences (TFR) – Swedish National Board for Industrial and Technical Development (NUTEK)	n.a.	Yes
Switzerland	– Federal Department of Interior (OFES/DEI) – Federal Department of Foreign Affairs (DOI/DFAE)	– Swiss National Foundation for the Encouragement of Research (FNS) – Paul Scherrer Institute (PSI)	9 research institutes for regional universities and federal polytechnics	Yes
Turkey		– Science and Technology Research Council of Turkey (TÜBITAK) – Atomic Energy Commission of Turkey	6 5 universities High Energy Physics Research and Development Centre	No (observer)
United Kingdom	– Office of Science and Technology	– Particle Physics and Astronomy Research Council (PPARC)	17 16 universities 1 support programme at Rutherford-Appleton Laboratory (RAL)	Yes
United States	– Department of Energy (DoE) – National Science Foundation (NSF)	– High Energy Physics Advisory Panel (HEPAP)	115 5 DoE national laboratories 110 universities	No

Table A.1. **Organisation of particle physics research in OECD countries and in Russia** (*cont'd*)

	Ministry in charge of particle physics	Councils or institutions	Nature of laboratories		CERN membership
Russia	– Ministry of Science and Technological Policy – Ministry of Atomic Energy	– Academy of Sciences – Joint Institute for Nuclear Research	21	Institute for High Energy Physics (IHEP) Institute for Theor. and Exp. Physics (ITEP) Budker Institute of Nuclear Physics Institute of Nuclear Research St. Petersburg Institute for Nuclear Physics Lebedev Institute of the Academy of Sciences Institute of Nuclear Physics of Moscow State University Moscow Physical Engineering Institute Institute of Applied Physics of Irkutsk State University 12 universities	No (observer)

1. Including fundamental nuclear physics.
n.a.: Not available.

Table A.2. **Consolidated annual budget for particle physics research (theory and experiments) in OECD countries and in Russia**

	Year of reference	(A) Salaries	(B) Basic support for laboratories	(C) Large national installations	(D) Annual financial support for experiments at international/foreign installations	(E) Total = (A) + (B) + (C) + (D) (in parentheses, million PPP¹ $)	(F) Contribution to CERN (million SF and in parentheses, million PPP¹ $)	Overall total = (E) + (F) (million PPP¹ $)	Remarks
			(Million national currency, except as noted)						
Australia (A$)	1994	2.8	0.5	0	0.6	3.9 (3.0)	0	3.0	
Austria (Sch)	1994	43.9	14	0	n.a.*	57.9 (4.02)	24.8 (11.2)	15.2	* Included in (B)
Belgium (BF)	1994	n.a.	36.9	0	8.9	n.a.	30.3 (13.6)	n.a.	
Canada (C$)	1993	n.a.	31.0	n.a.*	8.2	n.a.	0	n.a.	* Included in (B)
Denmark (DKr)	1994	8.4	3.6	0	0	12.0 (1.3)	17.6 (7.9)	9.2	
Finland (Mk)	1994	17.3	3.0	0	2.1	22.4 (3.7)	7.2 (3.2)	6.9	
France (FF)	1993	499*	170	0	67	736 (113.4)	[1994] 164.5 (74.1) [1993] 153.3 (69.1)	182.5	* Includes university professors and Ph.D. students
Germany (DM)	1994	165	99.5	267*	57.5	589 (274)	211 (98.1)	372	* Support for DESY
Greece (Dr)	1994	n.a.	n.a.	n.a.	n.a.	n.a.	3.75 (1.69)	n.a.	
Ireland (Ir£)	1994	n.a.	n.a.	n.a.	n.a.	n.a.	0	n.a.	

Table A.2. **Consolidated annual budget for particle physics research (theory and experiments) in OECD countries and in Russia** (cont'd)

	Year of reference	(A) Salaries	(B) Basic support for laboratories	(C) Large national installations	(D) Annual financial support for experiments at international/foreign installations	(E) Total = (A) + (B) + (C) + (D) (in parentheses, million PPP¹ $)	(F) Contribution to CERN (million SF and in parentheses, million PPP¹ $)	Overall total = (E) + (F) (million PPP¹ $)	Remarks
		(Million national currency, except as noted)							
Italy (Kilo L)	1994	71	13	101	43	228 (151)	141 (63.5)	214	
Japan (¥)	1993	2 770	15 000	n.a.*	1 500	19 270 (103)	0	103	* Included in (B)
Mexico (US$)	1993	1.7	1.6	0	0.2	3.6	0	3.6	
Netherlands (Gld)	1994	12.6*	3.4	0.8**	0***	16.8 (7.92)	43.8 (19.7)	27.6	* Does not include university professors and PhD students ** 4 M Gld (1994-98) *** 15 M Gld to be available for experimental equipment for LHC experiment
Norway (NKr)	1994	26	10	0	n.a.*	36 (3.9)	12.6 (5.7)	9.6	* Included in (B)
Portugal (Esc)	1994	n.a.	n.a.	n.a.	n.a.	n.a.	7.02 (3.2)	n.a.	
Spain (Ptas)	1994	2 250	150	0	250	2 650 (22.3)	68.9 (31.0)	53.3	

Table A.2. **Consolidated annual budget for particle physics research (theory and experiments) in OECD countries and in Russia** (cont'd)

	Year of reference	(A) Salaries	(B) Basic support for laboratories	(C) Large national installations	(D) Annual financial support for experiments at international/ foreign installations	(E) Total = (A) + (B) + (C) + (D) (in parentheses, million PPP¹ $)	(F) Contribution to CERN (million SF and in parentheses, million PPP¹ $)	Overall total = (E) + (F) (million PPP¹ $)	Remarks
		(Million national currency, except as noted)							
Sweden (SKr)	1994	n.a.	n.a.	n.a.	n.a.	n.a.	24.4 (11.0)	n.a.	
Switzerland (SF)	1993	25.9*	4	15.3	4.3	49.5 (22.3)	[1994] 38.7 (17.4) [1993] 36.3 (16.3)	38.6	* Includes particle and general physics teaching
Turkey (TL)	1994	n.a.	2 200	0	0.123*	n.a.	0	n.a.	* Grants to CERN in million SF
United Kingdom (£)	1995*	19.8	n.a.**	n.a.**	8.9	28.7 (46.3)	[1995] n.a. [1994] 127 (57.2)	n.a.	* Planning figures for 1995 ** Included in (D)
United States ($)	1994	n.a.*	486.4	n.a.*	1.9	668.3**	0	668.3	* Included in (B) ** Includes NSF's funds for universities
Russia (roubles)		n.a.	n.a.	n.a.	n.a.	n.a.	n.a.	n.a.	

1. Purchasing power parities (PPP): PPPs take into account the cost differences among countries for buying a comparable basket of goods and services including non-tradables. Therefore, PPPs give different results than the market exchange rate (MERS), which represents the relative value of currencies for goods and services traded across borders. As PPPs for 1994 and 1995 are not yet available, 1993 PPP figures are used for these two years. See OECD (1994), *Main Science and Technology Indicators, 1994-I*, Paris.
n.a.: Not available.

Table A.3. **Personnel employed in institutes/universities**

	Year of reference	Experimental physicists				Technical staff (including administrative staff, engineers, technicians)	Theoretical physicists				Remarks
		Permanent	Fixed-term	Students (PhD)	Total		Permanent	Fixed-term	Students (PhD)	Total	
Australia	1994	10	3	9	22	2	11	11	28	50	* Included in "Permanent"
Austria	1993	26	6	n.a.*	32	17	20	n.a.*	n.a.*	20	
Belgium	1994	13	4	n.a.	n.a.	38	8	6	n.a.	n.a.	
Canada	1993	64	40	60	164	n.a.	63	26	72	161	
Denmark	1994	7	6	1	14	3	7	4	3	14	
Finland	1994	15	14	12	41	18	10	13	11	34	
France	1993	356	30	108	494	969	115	10	33	158	
Germany	1994	454	148	221	823	1 358	86	24	42	152	
Greece	1988	10	0	5	15	6	5	2	0	7	
Ireland		n.a.	n.a.	n.a.	n.a.	n.a.	n.a.	n.a.	n.a.	n.a.	
Italy	1993	267	30	50	347	794	56	5	15	76	
Japan	1994	450	0	150	600	370	n.a.	n.a.	n.a.	n.a.	
Mexico	1994	6	1	15	22	30	34	6	34	74	* Excludes the physics theory division of NIKHEF
Netherlands	1994	36	9	30	75	69	2*	5*	7*	14*	
Norway	1994	25	n.a.*	16	41	7	14	n.a.	12	n.a.	* Included in "Permanent"
Portugal	1988	12	n.a.	16	n.a.	19	7	6	4	17	
Spain	1994	50	20	40	110	25	n.a.	n.a.	n.a.	n.a.	
Sweden	1991	26	9	24	59	23	16	11	29	56	
Switzerland	1993	27	96	75	198	91	8	21	19	48	
United Kingdom	1993	150	90	156	396	255*	89	69	171	329	* Excludes personnel recruited especially for a programme
United States	1990	1 322	344	767	2 433	n.a.	576	187	389	1 152	
Turkey	1994	8	5	6	19	n.a.	n.a.	n.a.	n.a.	n.a.	
Russia		n.a.	n.a.	n.a.	n.a.	n.a.	n.a.	n.a.	n.a.	n.a.	

Note: 1988 figures are taken from ECFA (1990), *The 1988 ECFA Survey of Particle Physics: Activities and Resources in the CERN Member States*, March.
n.a.: Not available.

Annex 2

Frontiers of Particle with Nuclear Physics

by

E.W. Otten
Institut für Physik, University Johannes Gutenberg, Mainz, Germany

When discussing applications and relations of particle physics to other sciences and technology, it is useful to touch upon the relation of particle physics to other physics disciplines as well. Its links to nuclear physics are naturally close, as particle physics emerged from nuclear physics some decades ago. While the two fields remained quite separate for some time, they now share some scientific interests and instrumentation (for an extensive treatment, see NuPECC, 1991). Main issues that enter the category of megascience are treated in this chapter.

1. Fundamental particles and interactions

One of the very central questions of physics has always been: What are the fundamental constituents of matter, and what are their interactions? During the first half of this century, physicists identified the atomic nucleus and its electronic cloud and discovered the laws of quantum mechanics, which determine the structure of atoms and molecules. They also discovered the constituents of the nucleus, protons and neutrons, and established the first models of nuclear structure. In the second half of the century, physics has proceeded to the next, still more elementary, level, the quarks that compose the nucleons. Figure A.1 schematises the hierarchy of physical systems and particles and indicates their typical dimensions.

In addition, high energy reactions have made it possible to observe the creation of other species of particles, including very heavy and unstable ones; they can be grouped into two categories, the quarks and the leptons, which differ, in particular, in the way they interact (see Figure A.2).

Just as important as these elementary particles are the quanta of the forces by which they interact – the gluons, the photons, the W and Z bosons, and the graviton. Present knowledge of these elementary objects and their fundamental interactions is contained in the so-called "Standard Model", which appears to give a satisfactory description of all the observed phenomena.

What are the burning questions, and how does nuclear physics contribute to solving them? What follows is an attempt to identify some of them and illustrate, with some concrete examples, the interplay between nuclear and particle physics and their common experimental frontiers.

173

Figure A.1. **A simplified view of the subatomic world**

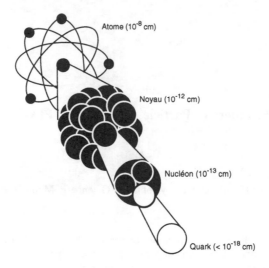

Source: Nuclear Physics European Collaboration Committee (NuPECC), 1991.

Figure A.2. **The Standard Model**

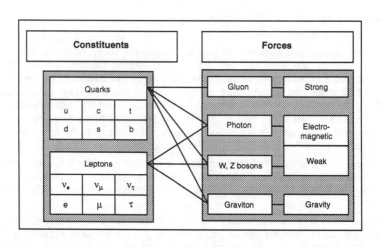

Note: The fundamental constituents are indicated on the left, the forces on the right and interconnections between forces and particles are indicated by the lines.
Source: Nuclear Physics European Collaboration Committee (NuPECC), 1991.

2. Examining the Standard Model and beyond

Physicists are convinced that the Standard Model is not the ultimate description of nature, because its structure is complicated and it contains many arbitrary and unrelated constants. How can we test the limits of its validity, and how do we get any experimental hint of the physics that lies beyond? What are the experimental signatures that we should seek?

Neutrino masses and oscillations

Although the Standard Model generally assumes that neutrinos have zero masses, there is no fundamental reason or need for this. The measurement of non-zero masses for one or more of the three generations of neutrinos – ν_e, ν_μ, ν_τ – would give a clear signal that the model is incomplete and a means of selecting among theories beyond the Standard Model. Many experiments with nuclei have been mounted in laboratories around the world to look for evidence of a non-zero mass for the electron antineutrino $\bar{\nu}_e$ in the beta-decay spectrum. The current best results are consistent with zero mass, with an upper limit of 7 eV/c². Since the energy equivalent of this limit is already comparable to molecular excitation energies accompanying the beta decay, these experiments also touch upon central problems in present atomic and molecular physics.

Over the years, several experiments have successfully detected the flux of solar neutrinos, and to some extent also their energy distribution, in large, well-shielded underground neutrino detectors [Homestake (United States), Kamiokande (Japan), Gran Sasso (Italy), Bakhsan (former Soviet Union)]. When compared to astrophysical calculations of neutrino production in the sun, these measurements have revealed an insufficient neutrino flux, especially on the high energy side of the spectrum. One might speculate that at least one type of neutrino has a very small but still non-zero mass, of the order of a few times 10^{-3} eV/c², and mixes with another neutrino generation so that the flavour of the neutrino changes as it oscillates to the other type. Extensive terrestrial measurements of neutrino oscillations with reactor- and accelerator-produced neutrinos have led to results consistent with zero mass, within limits that can roughly be characterised by the inequality $\Delta\, m^2 \cdot \sin^2 (2\theta) < 10^{-2}$ eV²/c⁴, for the product of the difference of the squared masses m² and the fraction of mixing $\sin^2 (2\theta)$.

A variety of solar neutrino and oscillation detectors, which are pushing the techniques of nuclear radiation detectors to unprecedented limits, have come or will be coming into operation.

Double beta decay

Many theories beyond the Standard Model assume that neutrinos are Majorana particles, *i.e.* that they do not distinguish between being particles and antiparticles. These theories also include mechanisms that give mass to neutrinos. Such properties would lead to a neutrinoless double beta decay of some nuclei, with only two electrons emitted. Extremely sensitive and clean detection methods have been developed in the search for such events. At present, the best upper limit of such a decay rate is 4×10^{-25}/year, obtained for ^{76}Ge; this number can be converted into an upper limit of the neutrino mass of 1.1 eV/c². All these experiments on the nature of neutrinos involve particular problems of nuclear structure and reactions, such as the knowledge of various transition matrix elements, which must be thoroughly understood in order to do a quantitative analysis.

175

Neutrinoless muon decay

The occurrence of neutrino oscillations and neutrinoless double beta decay would prove that neither the flavour nor the number of neutrinos are conserved quantum numbers. The search for a neutrinoless decay of the muon into an electron raises the same question for the charged leptons. In extensive measurements at the meson factories at the Paul Scherrer Institute (PSI) (Switzerland), TRIUMF (Canada), and LAMPF (United States) no such event has been found in a total number of about 10^{12} muon decays.

Free neutron decay

The interdisciplinary role played by neutron beams furnished by reactors touches not only the whole of the life and the material sciences but also fundamental questions of quantum mechanics and particle physics. In particular, neutron beta decay is a unique source for probing with high precision the semi-leptonic sector of the weak interaction. Tests by neutron decay of the two components of different symmetry of the weak Lagrangian interactions, the so-called vector and axial-vector coupling constants, and of the unitarity of the quark mixing matrix, which is a measure of whether the set of the six quarks participating in the weak interactions is complete, reach a level of accuracy of 10^{-3}.

Facilities and perspectives

The research activities touched upon in this section range from university-based experiments of usual size up to huge detectors that fill spacious halls in underground laboratories, ideal shelters against any disturbing background radiation. The latter almost reach the level of megascience.

Consequently, international collaboration is developing. Italian and US groups are sharing the costs for the MACRO detector at the Gran Sasso laboratory. It has a volume of about 4 000 m^3, in which all kinds of cosmic rays are measured and searches are made for magnetic monopoles, hypothetical particles which would carry an as yet never observed free magnetic charge, by analogy with the electric charge. In the LVD detector, also at Gran Sasso, 1 400 tonnes of liquid scintillator observe neutrinos from stellar collapses. Canadian and US physicists are joining efforts to build the Solar Neutrino Observatory (SNO).

International collaboration is also driven quite naturally by the aim, even the need, to pool the best expertise and facilities, wherever they are, in order to realise a complex project. The Gallex collaboration on solar neutrinos and the double beta decay experiment on ^{76}Ge are good examples: Russia provided the otherwise unattainable material, namely the separated isotope ^{76}Ge; the United States and Germany contributed the highly specialised nuclear chemistry and detector technology, respectively; the neutrino source for calibrating the Gallex detector was bred in a French reactor; and Italy hosted the experiments in the unique Gran Sasso laboratory.

Terrestrial neutrino oscillation experiments in the low energy regime are bound to power reactors (*e.g.* at Bugey, France, and Goesgen, Switzerland) and to accelerators in the high energy regime. Their size and complexity nowadays equal those of typical high energy experiments. Hence, they are realised in international collaborations. New dimensions in long distance experiments may be reached, *e.g.* with the source at CERN and the detector at Gran Sasso.

For the large underground experiments, which attract particle and nuclear physicists, as well as astrophysicists, NuPECC explicitly recommends international co-ordination and collaboration in the 1991 report mentioned above.

3. Symmetry tests with atoms and nuclei

In weak interaction, nature provided us with an example of breaking a fundamental symmetry, namely the mirror symmetry (parity) which states that any physical process and its mirror image should obey the same laws. Since 1957, when this parity principle was found violated in nuclear beta decay, symmetry tests have been a basic tool for investigating fundamental interactions. Nuclear and atomic physics provide precise and powerful tests of these symmetries with far-reaching consequences.

The question of whether weak interactions are exclusively left-handed remains open and has been investigated in many scenarios. Some of them, which analyse nuclear beta decay experiments with a high degree of precision, open up possibilities of discovering right-handed components.

Thirty years was discovered a rare decay mode of a particular particle, the K^0 meson, which revealed that the weak interaction breaks not only parity but also, and simultaneously, the symmetry between particle and antiparticle and the symmetry with respect to the reversal of the sense of time. Although this phenomenon is so weak as to be almost unobservable in any other process, it still provides our only key for understanding why no antimatter seems to be left over in the Universe from the time of the Big Bang. Astrophysical data indicate, indeed, that the Big Bang initially produced a very few more particles than antiparticles – about one out of one billion. This tiny surplus has then finally survived mutual annihilation. One concludes this from the fact that the number of photons resulting from this mutual annihilation exceeds the number of protons in the Universe precisely by a factor of about a billion.

The weak time-reversal violating force should manifest itself also by the electric dipole moment of an elementary object such as a neutron, a nucleus, or an atom. Very subtle experiments have tested all these systems and set stringent bounds on many model calculations. Recent measurements on the free neutron have shown that its electric dipole moment is smaller than what would result from separating a pair of elementary charges by 10^{-12} of the neutron's diameter. Measurements of directional correlations in the decay and reactions of polarised neutrons and nuclei are another rich source of information on the violation of time reversal symmetry.

The idea of using atoms and nuclei as "laboratories" for investigating fundamental interactions holds only to the extent that these laboratories, including all of their corners, are really known and understood by experiment and theory.

Facilities and perspectives

Symmetry tests are still performed quite often in tabletop experiments, but in order to go beyond present limits, all the fascinating, sophisticated tricks of modern quantum optics and atomic physics are needed. Insofar as they make use of neutron and muon beams, their future depends strictly on the availability and further development of high flux reactors, spallation sources and meson factories, large facilities that run broad multi-disciplinary research programmes. The perspectives in this area are reviewed in this volume (and in OECD, 1994).

4. Hadrons as quark systems and nuclear constituents

The hadrons are the essential link between particle and nuclear physics. They are the simplest and so far the only structures in which quarks manifest themselves. Quark-antiquark pairs (mesons) or three quark systems (baryons) are the simplest ways to form the colour singlets demanded by the theory of strong interactions, the quantum chromodynamics (QCD). Some models inspired by the as yet unsolvable QCD suggest the existence of more complex colour singlets, formed by more than three quarks, by gluons alone, or as a hybrid consisting of a certain number of quarks and gluons. Although such objects have been sought intensively in hadronic reactions at medium energies, for instance in proton-antiproton annihilation experiments where they are expected to form, there are only a few uncertain candidates.

The QCD problem of hadronic structure

It is a euphemism to talk about hadrons as simple systems. It is only their quantum numbers which are simple, for instance, the spin or the number of valence quarks contained. In addition, they exchange the virtual gluons binding them and carry a sea of virtual quark-antiquark pairs.

All this gives rise to a complex structure, to which we have still very limited access by theory. Why is this so? The problem is that the QCD theory is manageable so far only at very high energies where the strong coupling constant becomes small. But it leads into yet unsolved mathematical problems in the lower energy regime because the basic QCD colour force is much stronger there. Therefore, the physics of hadrons and nuclei operates so far with phenomenological models of effective forces and effective particle masses.

However, experimental physics offers a great variety of investigations into the secrets of hadronic structure. One means of access is to study reactions at medium energies where the internal degrees of freedom of the nucleons are excited. A very fruitful way to study the structure of the nucleon is to scatter electromagnetic or weakly interacting probes – electrons, muons, photons and neutrinos. Since their interaction is well understood at all energies, they give reliable testimony of what they have seen when passing through the nucleons or nuclei. It was in this way that the quarks were discovered in the late 1960s as the basic constituents of the nucleons which remain confined in them at every energy. Further breakthroughs in this area require measurements at very short time scales and distances. The shortest distance between the locus of a reaction and its detection that nature offers is the distance between neighbouring nucleons in a nucleus, one serving as target, the other as detector by means of a secondary reaction. Using the "nucleus as a laboratory" makes it possible to study the interaction of knocked-out quarks with matter.

Nucleons and nuclear binding

The strong interaction does not exhaust its force by binding quarks to hadrons. A residual force is left over which binds nucleons to nuclei. But it is no longer governed by the exchange of the coloured gluons. Colour remains confined in the nucleons as the colour potential rises steeply with distance. Instead, except at very small distances, nucleons exchange uncoloured pions and other quark-antiquark systems, which is energetically more favourable.

How do the nucleons respond to this binding? There is evidence that their magnetic moment changes in the nuclear surrounding and that the internal momenta of their individual quarks are reduced in the nuclear medium. The influence of the nuclear binding on the nucleonic structure provides valuable information for understanding the nucleon in terms of QCD. At that point, the open questions of nuclear and particle physics are inseparable.

Facilities and perspectives

Physicists are convinced that very detailed and careful experiments are needed to clear up these complex problems. Present activities still exploit the electron beam of SLAC (polarised and unpolarised) as well as muon beams at CERN and Fermilab. The Hermes project at DESY will scatter the polarised, circulating 30 GeV electron beam of HERA from polarised internal H and ^{3}He targets. At lower energies, experimenters use the electron beams from the Mainz Microtron (Germany, 850 MeV) as well as from stretcher rings at NIKHEF (Netherlands, 700 MeV), at Bonn (Germany, 3.5 GeV), and at MIT (United States, 800 MeV, soon to come). Many of these experiments exploit the polarisation observable in electromagnetic as well as in electroweak reactions.

In particular, there is a desire for electron accelerators delivering a continuous beam (100 per cent duty cycle) that makes it possible to detect the scattered electron simultaneously with the reaction products in a so-called exclusive experiment. The relevant energy scale ranges from about 1 GeV, where nucleon resonances and form factors dominate the scene, up to about 30 GeV, where the quark picture rules. At the lower end of this scale, the above-mentioned facilities are more or less meeting the experimental demands for duty cycle and intensity. With increasing energy, increased intensity is needed; for example, in the case of coherent scattering from the three quarks in a nucleon, the cross-section falls off like $1/E^4$. A beam intensity of some 10μ A on a target of thickness 1 g/cm^2 could provide the required luminosity of the order of $L = 10^{37}$ cm^{-2} s^{-1}.

Although the energy scale is moderate by present high energy physics standards, the high demands on the beam characteristics drive such facilities into the "megascience" range. These requirements can be met only by superconducting accelerator structures at energies above several GeV.

The first, and so far only, continuous electron accelerator entering the high energy/high intensity scenario will be CEBAF in the United States. This superconducting linac, through which the beam is recirculated five times, is designed for an energy of 4 GeV with possible upgrade later to 6 GeV. It will become operational by the end of 1994. Its total cost, including the main experiments (two high resolution magnetic spectrometers in coincidence, one toroidal, superconducting magnetic spectrometer of high acceptance, one high momentum electron spectrometer in coincidence with a general purpose hadron spectrometer), amounts to $620 million. It will serve a community of about 500 scientists presently, including contributing groups from all over the world, although it is a national facility. With regard to future plans, a CEBAF workshop has discussed an upgrade of the facility into the range of 8 to 12 GeV.

Starting from a French initiative, European nuclear physicists are promoting the ELFE project. This superconducting linac will deliver a continuous electron beam in the upper half of the desired energy range between 15 and 30 GeV which forms the bridge from the non-perturbative low energy to the perturbative high energy regime of QCD. The first extensive project report on ELFE (Arvieux and de Sanctis, 1993) counts about 30 scientific proposals and outlines the accelerator concept as well as the experimental facilities (a pair of magnetic spectrometers, a forward angle spectrometer, a di-muon spectrometer, and a multipurpose 4π detector). Based on present technology, the cost estimate for the first 15 GeV stage of the project, not including salaries and experiments, amounts to FF 1 760 million.

Accelerator physicists from Saclay (France) have joined the international "Tesla Test Facility" collaboration which performs extensive R&D work aimed at increasing the performance and decreasing the costs of superconducting accelerator structures. It is generally felt that progress in this area is needed if ELFE is to meet its final goal of 30 GeV. This R&D work will also bring about important technological steps, such as alternatives to synchrotron and laser radiation at short wave length, etc.

5. Quark-gluon plasma

How does nuclear matter behave at very high energy, where the density and temperature of its constituents increase far beyond their usual value in a nucleus? Theoretical investigations have concluded that a new phase of matter should exist under these conditions, where the quarks and gluons are finally freed from their confinement within nucleons, pions or other hadrons. They may form instead a dense ensemble of freely propagating and interacting quarks and gluons. This state of fully dissociated nucleons is called the "quark-gluon plasma" and is characterised by a phase transition in which the chiral symmetry of QCD is globally restored. This symmetry respects the property of QCD that relativistic quarks interact only with those of same chirality, a parameter that characterises certain correlations between the directions of spin and momentum.

Collisions between heavy nuclei at extremely high energies (typically many TeV in the centre-of-mass system) make it possible to investigate this fundamental question. During such collisions a good fraction of the kinetic energy is used up for heating, compressing and exciting constituents to conditions where this ultimate phase of matter should form, at least for a very short period of time. With the production of the quark gluon plasma we can hope to simulate on the laboratory scale a certain state of matter that we believe to have existed as a transient phase of the Universe during the Big Bang.

From these events, thousands of secondary particles will originate, which have to be analysed in the search for the signature of the quark-gluon plasma. Their extraordinary complexity is typical of modern particle and nuclear physics; if the technology of particle detection, data acquisition, and computing are to achieve new dimensions at a reasonable cost level, an effort, which will necessarily create high technology spinoffs, will have to be made.

Facilities and perspectives

Energetic heavy ion collisions which heat and compress nuclear matter far above its ordinary value are presently performed at the GSI (Germany), at the AGS at Brookhaven (United States), and at CERN, with beam energy of roughly 1, 12, and 200 GeV/nucleon, respectively, hitting fixed targets. In principle, beams of any element up to the very heavy ones can be made available. Although these experiments lead into yet unexplored regions, it seems unlikely that they will provide enough centre-of-mass energy to reach the transition to the phase of the quark-gluon plasma. Such hopes concentrate, therefore, on future heavy ion collider experiments.

RHIC at Brookhaven (United States) will realise collisions at a centre-of-mass energy of 2×100 GeV per nucleon in a collider ring equipped with superconducting bending magnets. The collider will serve two major experiments, STAR and PHENIX. The former is already under construction, and the technical proposal for the latter is being finalised in 1994. At present a community of about 800 users is engaged in RHIC. Depending on the speed of funding, RHIC will become operational between 1997 and 1999.

A still larger centre-of-mass energy is envisaged in the LHC project at CERN. Two lead beams would collide at a total energy of about 2×600 TeV. About 300 physicists from 42 institutions are engaged at present in establishing the technical proposal for the only experiment foreseen, the ALICE detector. Its building cost should be about SF 100 million. Under present planning, the construction of ALICE should be finished at the same time as the LHC, in 2003.

6. Advisory committees in nuclear physics

CEBAF and RHIC are by far the largest nuclear physics projects being realised at present in the United States. Clearly, the communities from which they originated represent only some of the US nuclear physicists. Therefore, the American Nuclear Science Advisory Committee (NSAC) took the task of:

- examining the projects;
- promoting them in order to create a consensus regarding their priority;
- fighting for approval by government.

The outcome certainly justifies NSAC's role. Today the programmes at CEBAF, AGS, and RHIC are attracting groups from all over the country and abroad, among them a good share of particle physicists.

Decisions in nuclear science politics traditionally have been taken at the level of national committees and agencies. However, with the possible exception of the United States, no single nation will be able to afford the human and capital investment demanded by the upcoming generation of facilities. Hence, international co-operation is necessary.

In parallel to the CEBAF proposal, a French community proposed a similar project for France, which was reviewed in 1989 by a committee of the French Academy of Sciences. Its advice was twofold:

- to head for higher energies;
- to seek European collaboration.

The ELFE project was born.

Soon after, in 1990, an initiative of a group of European nuclear physicists resulted in the foundation of the Nuclear Physics European Collaboration Committee (NuPECC) as an associate committee of the European Science Foundation (ESF). The committee's terms of reference state that its tasks are:

- to provide a forum for the discussion of the provision of future facilities and instrumentation;
- give advice, in general, on the development of nuclear physics.

The 21 delegates to NuPECC from 14 western European countries meet about three times a year. NuPECC's primary public activities are:

- publishing a quarterly journal, *Nuclear Physics News*, together with the European Physical Society;
- organising European workshops on the status and future development of the field which result in NuPECC reports.

NuPECC's above-mentioned report contains three recommendations of interest in the context of particle physics and large international facilities. Summarised, they are as follows:

- Regarding low energy aspects of fundamental interactions and particle properties, steps shall be taken to ensure that access to adequate beams of neutrinos, muons, and neutrons is available and to facilitate co-ordination of large underground experiments on a European scale.
- Regarding continuous electron beams, a European project group is proposed, charged with identifying the key experiments, determining the initial energy and optimal design of the accelerator, and establishing instrumentation for the initial experimental programme. The above-mentioned ELFE report responds to this recommendation. NuPECC will issue a

second statement on ELFE soon, encouraging the proponents to continue and in particular to reinforce technical R&D on the accelerator, to develop the physics methods, to work out potential applications, and to create an enlarged community.
- Regarding heavy ion collisions at very high energies, the NuPECC report recommends making full use of the lead beams soon to be available at the SPS at CERN. It also recognises the new perspectives offered by plans to run colliding lead beams in the LHC.

Conclusion

In conclusion, it should be underlined that the communities involved in nuclear physics at high energies as well as the accompanying advisory committees are very much aware of the scientific and practical advantages as well as the need to build up common frontiers with particle physics.

References

Arvieux, J. and E. de Sanctis (eds.) (1993), *Italian Physics Society Conference Proceedings*, Vol. 44.

Nuclear Physics European Collaboration Committee (NuPECC) (1991), *Nuclear Physics in Europe: Opportunities and Perspectives*, November.

OECD (1994), *Neutron Beams and Synchrotron Radiation Sources*, Paris.

Megaprojects in the Field of Particle Astrophysics

by

M. Spiro and L. Sulak*

C.E.A. Saclay, Gif-sur-Yvette, France

Introduction

For the past two decades particle astrophysics has developed into a new field at the intersection of particle physics and astronomy. Many experiments are currently being carried out. From the present status of these experiments, we can recognise trends and anticipate the emergence of megaprojects which should be initiated soon to operate in the early decades of the twenty-first century.

We do not consider space astronomy experiments since they have already been a topic of an earlier Megascience Forum study. We do not review all of particle astrophysics. Indeed, this field covers a wide variety of experiments, including many small-scale projects. For example, the search for dark matter, an extremely rich field of research, consists of many relatively small experiments with specific goals. Although some international co-ordination may be appropriate, we do not consider this field as belonging to megascience. We do discuss, in some depth, gravitational waves, high energy cosmic rays, high energy neutrinos, monopoles and proton decay, and low energy solar neutrino detectors. This physics now requires experiments that approach the scale of megascience. However, since the experiments do not fall into the traditional disciplines of astronomy or of accelerator physics, they tend to have the character of orphans.

Without the advocacy of an international organisation similar to CERN (European Laboratory for Particle Physics), ICFA (the International Committee for Future Accelerators), or ECFA (the European Committee for Future Accelerators), it is unlikely that the next generation of worthy astrophysical projects, on the fringes of particle physics, will ever be realised. We call for a broad international sponsorship of particle astrophysics to be developed in the near future.

1. Gravitational-wave detectors

According to general relativity, a gravitational field induces a curvature of space. This curvature is associated with a change of the metric or distance scales. Gravitational waves (time-

---

* Also on leave from Boston University, Boston, MA, United States.

dependent metrics) are expected to propagate, due to a strong variation of the gravitational field. These variations must be non-axisymmetric (quadrupole in nature) to produce waves. These waves can be characterised by the amplitude of the change with time of local distances and by their frequencies. The amplitude depends both on the nature of the source and its distance from the detector. The frequencies can vary over a wide range, depending on the nature of the phenomenon. The sensitivity of a given detector can then be parameterised by two figures: the sensitivity to distance changes and the frequency bandwidth.

The detectors presently operating are resonant bar detectors (1 000 Hz frequency) at very low temperature (Weber bar type). The length of a bar is monitored at the 10^{-20} level with a bandwidth of around 30 Hz, depending on the signal/noise ratio. Such bars are being operated in the United States (Maryland, Louisiana), in Europe (Frascati, Rome, CERN, Meudon), in Russia (Moscow), in Australia (Perth), in Japan (Tokyo), and in China (Guangzhou).

Laser interferometers several kilometres long should make it possible to reach a much better sensitivity, about 10^{-22}, in a bandwidth from 10 Hz to 1 000 Hz. This would open a new window in astronomy, as it would put many new phenomena within reach.

The best understood of potential gravitational-wave sources are coalescing compact binary systems composed of neutron stars and black holes (*i.e.* two neutron stars, or one neutron star and one black hole, or two black holes). The famous binary pulsar PSR 1913 + 16 is an example which could be detected as the two objects finally merge into either a big neutron star or a black hole. A detailed study of this binary system has shown that the two objects are losing orbital energy and gradually spiraling inward. (This work received the 1993 Nobel prize.) They will merge in about 10^8 years.

Calculations of the energy loss rate of coalescing binary objects show that this energy is radiated in the form of gravitational waves. However, these gravitational waves will not be detectable from Earth until the last five minutes during which the two objects collide and merge. A detailed analysis of the gravitational wave form of such an event will tell us about the equation of state of neutron stars and about disruption effects during the collapse. More generally, it will reveal the behaviour of matter under the most extreme conditions, just before it disappears into a black hole.

The search for a coincidence between gravitational-wave pulses and γ-ray bursts might give a clue to the origin of the puzzling and unexplained isotropic point-like γ-ray events. Failing or succeeding in observing these coincidences might make it possible to rule out or prove the cosmological versus galactic origin of γ-ray bursts. Indeed, the anticipated sensitivity of the new gravitational wave antennas should allow detection of coalescing stars as far away as 200 Mpc at an expected rate of a few per year. Figure A.3 compares the projected interferometer sensitivities with the strength of the waves in spiral of compact binaries during the last few minutes.

The new antennas should make it possible to detect supernovae if the collapsing core becomes non-axisymmetric during its rapid rotation. Periodic waves from spinning neutron stars in our galaxy, if they are not axially symmetric, should be observable. A stochastic gravitational-wave background from the early Universe would be visible if its energy density, in the bandwidth of 40 to 120 Hz, is greater than 10^{-10} of that required to close the Universe. This is predicted both by the theory of cosmic strings or more generally by inflationary universe scenarios, coupled with the COBE results on the large-scale anisotropy of the microwave background. This stochastic gravitational signal can only be identified if it is coherent on at least three antennas separated by thousands of kilometres.

Three laser interferometer gravitational wave detectors several kilometres long are being constructed at three sites: Hanford (Washington) and Livingston (Louisiana) in the United States (LIGO project) and Pisa in Italy (VIRGO project). Although LIGO and VIRGO have been planned

Figure A.3. Projected interferometer sensitivities

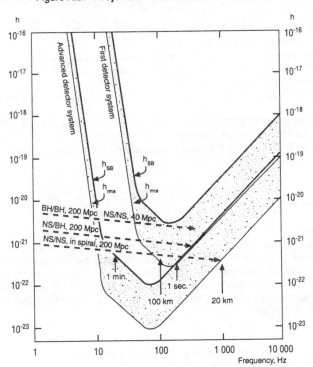

Note: h is the relative length scale changes. BH stands for black hole, NS stands for neutron star.
Source: Authors.

and funded independently, these three cross-correlated detectors will be a world baseline experiment. All three interferometers are necessary to extract the full information carried by gravitational waves: the direction of the source, the energy carried by the gravitational waves, and the exact waveform as a function of time. The planned sensitivities of VIRGO/LIGO as a function of frequency are shown in Figure A.3. They are compared with the expected signal from the collapse of a binary system with a black hole (BH) or a neutron star (NS).

LIGO/VIRGO might well reveal surprises about the Universe or about gravity that cannot be learned in any other way. The LIGO-VIRGO gravitational wave detectors should be operational around the year 2000. The project is funded at a total cost of around $400 million and may attract about 200 physicists.

2. High energy gamma, proton, and heavy nuclei astronomy

High energy cosmic rays – γ-rays and charged particles – have always been a subject of common interest for astrophysicists and particle physicists. Indeed, the founding generation of particle physicists had their training in cosmic-ray physics. Various questions maintain this common interest:

- Where are the sites of acceleration?
- What are the acceleration mechanisms?
- What is the chemical composition (gamma, proton, or heavy nuclei)?
- What are the energy spectra, are there cut-off energies, and why?
- Do we understand the interactions of the highest energy particles?
- Are there particles in the cosmic rays which have not been found at present accelerators?

Presently, a wide variety of experiments can be classified as taking place in the high, very high, ultra high, and extreme energy regimes, each of which has an applicable detection technique. Experiments below 20 GeV fall in the domain of satellite experiments. The range of experiments considered in this report are those above.

Below 20 GeV (satellite experiments)

Below 20 GeV, the rate of particles impinging on the Earth's atmosphere is high enough to be detected by satellite experiments. For instance, the Compton Gamma Ray Observatory satellite, with its four instruments covering the 30 keV – 30 GeV energy range, has already provided many results, including some on γ-ray bursts. A total of six pulsars (galactic sources) have been identified, and more than 20 active galactic nuclei (extragalactic sources) have been observed to be γ-ray emitters. The statistics collected by the small sensitive area of satellites limit their energy domain at an upper value of 10-20 GeV.

From 20 GeV to 200 GeV γ-ray astronomy (Cerenkov technique)

At this time, this is a region where no observations are possible up to the present 200-400 GeV threshold of ground-based atmospheric Cerenkov telescopes, which use the atmosphere as a giant calorimeter by observing the Cerenkov light radiated by the showers in the atmosphere. To fill the gap between 20 and 200 GeV, it has been proposed to collect the Cerenkov light from low energy gamma showers by many individual mirrors (e.g. the heliostats of a solar power plant, Themis, in France) focused onto a single light detector.

From 200 GeV to 10 TeV γ-ray astronomy (Cerenkov technique)

Above 200 GeV, reaching into the TeV region, is the domain of very high energy astronomy, which has been explored for many years by relatively simple detectors. One galactic source, the Crab Nebula, has been identified (by the Whipple collaboration in the United States, by ASGAT, and by the Themistocle collaboration in the French Pyrenees). A pulsar also has been observed in the Southern Hemisphere (PSR 1706-44) by a Japanese group. Detailed studies of the energy spectra of such sources should shed light on the acceleration mechanisms of cosmic rays in neutron stars.

Whipple's observation, two years ago, of the extragalactic source Mrk 421 in the TeV region has opened the possibility of studying the acceleration mechanisms in active galactic nuclei and of probing the absorption due to gamma-gamma interactions with the diffuse infrared background radiation.

An important goal, which could also be reached in this energy range, is the discovery of the annihilation products of dark matter coming either from the halo or from the center of our Galaxy. In one scenario, the lightest supersymmetric (SUSY) particle dominates the dark matter of the galactic halo. Future detectors may be sensitive to the annihilation of SUSY particles into two gammas near the galactic center. The signature would be a monoenergetic γ-ray line around the mass of the lightest supersymmetric particle. A strong enhancement of radiation would be observed towards the centre of our Galaxy. Calculations show that a detector of 45 000 m^2 is required to get a significant signal-to-noise ratio.

From 50 TeV to 100 PeV astronomy (scintillation arrays)

At higher energies (ultra high energy, from around 100 TeV to 100 PeV, 1 PeV = 10^{15} eV), where charged particle showers can penetrate the atmosphere and reach the ground, it is possible to measure directly the direction (by timing) and the energy (by pulse height) of air showers with arrays of particle detectors, usually scintillators. Due to the higher energy threshold of the detectors, these devices have to deal with smaller fluxes but compensate by enlarging the surface of the array. Very large arrays have been recently put into operation. A vast quantity of statistics has been accumulated by HEGRA in the Canary Islands and by Cygnus I and Casa Mia in the United States. These three large arrays have now been joined by a fourth in Tibet which compensates for its smaller size by its very high altitude. These telescopes are all dense arrays of charged particles detectors.

In this region no stable gamma sources have been observed. However, the energy spectrum of the primary cosmic rays shows a knee around 10 PeV. The flux dN/dE is proportional to $E^{-2.7}$ below 10^{16} eV but proportional to E^{-3} for E above it. The reason for the knee is unknown. It might be associated with a change of the cosmic ray composition from primarily hydrogen (H) to an enrichment in heavy nuclei (Fe). For a given energy primary, H has a higher magnetic rigidity and can therefore leak out of the galactic magnetic field more easily than a heavy primary.

Above 10 PeV

Beyond 10^{16} eV (the extreme energy range), two detection techniques are used: air scintillation (the Fly's Eye technique) and giant arrays of sparce detectors, as at Akeno in Japan, Yakutsk in Russia, Sydney in Australia, and Haverah Park in the United Kingdom. No clear γ-ray signal has been observed, but important results on charged hadrons (the primary cosmic rays) have been reported. A flattening of the energy spectrum around $10^{18} - 10^{19}$ eV (the "ankle") has been observed by the Fly's Eye collaboration in the United States. It is probably associated with a change from a predominantly heavy to a predominantly light composition.

The remaining questions are: Is there a galactic disk excess, are there point sources, and what is the exact energy cut-off? A cut-off is expected because of the interactions of extreme energy protons or heavy nuclei with the photons of the primordial microwave background (pion photoproduction). The energy of the cut-off depends on the chemical nature (proton or heavy nucleus) of the extreme energy cosmic rays and on their origin (galactic or extragalactic). At the extreme energies, neutrons can reach the Earth from everywhere in the local group without decaying; a neutron with 10^{20} eV energy has a mean decay length of 1 Mpc. This opens up the possibility of a new source of interactions in the atmosphere. To get sufficient statistics to sort out the different scenarios, detectors with an area of 10 000 km^2 are needed.

The study of cosmic rays – their energy spectrum, composition, and searches for point sources or preferred sites of acceleration – uses a wide variety of techniques (see Figure A.4) and attracts a

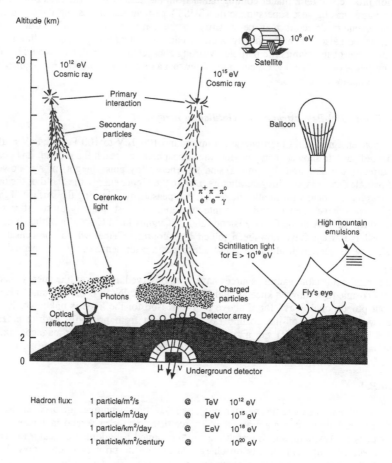

Hadron flux:

1 particle/m²/s	@	TeV	10^{12} eV	
1 particle/m²/day	@	PeV	10^{15} eV	
1 particle/km²/day	@	EeV	10^{18} eV	
1 particle/km²/century	@		10^{20} eV	

Source: Authors.

large community (several hundred physicists) of relatively small groups throughout the world. The physics goals are promising and achievable.

Our recommendation is to co-ordinate and focus these efforts towards two major universal detectors (one in each hemisphere, with some overlap between them). The total size of these arrays should be around 10 000 km². They could be built like Russian dolls: starting from an ultra-dense core (*e.g.* a "solar plant" array) to cover the 20-200 GeV range above satellite detector sensitivities, then a less dense 100 000 m² air Cerenkov telescope for the 200 GeV – 10 TeV range, surrounded by a 10 km² charged particle or Cerenkov array for the 10 TeV – 10 PeV range, and finally a sparce array combined with a Fly's Eye detector to cover the extreme high energy range.

The choices of the techniques would be debated by the international community with a view to optimising sensitivity. Overlaps in the energy ranges covered by the different techniques would allow cross-calibrations. Using the same shower events would reduce the systematic errors in the energy and angular resolutions. The total cost of this facility might reach the $200 million level. About 400 physicists would participate in the project.

However, it must be recognised that it is not yet clear that the two most ambitious goals of the high energy cosmic ray community – the search for point-like γ-ray sources from 20 GeV to 10 TeV and the study of the cut-off at extremely high energies (10^{20} eV) for protons or heavy nuclei – can be realised in one detector. Technical considerations, *e.g.* the ''seeing'' at the site and the staging of the techniques and collaborations, may prevent the joint deployment of these two searches.

3.   High energy neutrino, monopole and proton decay facilities

Underground cosmic ray detectors

Over the past 15 years a group of large, second-generation underground detectors have made significant contributions to our understanding of basic particle physics as well as neutrino astrophysics. With their large observed mass, these detectors probe very rare processes, such as the decay of the proton. This process has not been active since the ultra high energies of the Big Bang, when the time-reversed reaction created the excess of matter over antimatter in the Universe. The study of nucleon decay holds the key to unravelling the theory of particles and forces at ultra high energies. There, the strong and the electroweak interactions were one unified force.

With their large mass and low sensitivity, the underground detectors also can see low energy neutrinos from supernovae and, to some extent, from the Sun. They can verify our understanding of the processes of energy generation in the Sun and the generation of heavier elements from lighter elements by fusion reactions in a supernova. With their large area, these detectors can register upward-going muons (produced by cosmic ray neutrinos after traversing the Earth) and point back to the source of the production of their parent neutrinos, neutrinos that, unlike the photons received by light telescopes, escape from active galactic nuclei, black holes, etc., unabsorbed by intervening gas and clouds. With large area and high sensitivity, these detectors can also observe the passage of a magnetic monopole, a conjectured fundamental magnetic particle left over from the Big Bang.

By bringing the large-scale techniques of accelerator physics to the study of cosmic ray neutrinos, monopoles, and the search for nucleon instability, these detectors have established the following fundamental results:

1. Grand unification of the strong and the electroweak force does not occur in the simplest model of unification. The rate of proton decay is at least four orders of magnitude too slow. On the other hand, this limit on the rate is consistent with the next most likely extension of the theory, supersymmetry. These investigations probe an energy scale of 10^{12} TeV, far beyond that of any accelerator we can ever hope to build (the energy of the proposed LHC is 7 TeV). These extreme energies have not existed since the first few instants after the Big Bang.

2. By observing neutrinos from a supernova (1987A), the standard model of the final gravitational stellar collapse into a neutron star, predicted by Chandrasekhar in the 1930s, has been confirmed for the first time. Further, the model of the production of all elements heavier than iron has been checked. By comparing the arrival times of the neutrinos underground with the light seen hours later in telescopes on the surface, upper limits on the mass,

magnetic moment, and electric charge of the neutrino have been set. The equivalence principle of general relativity for weakly interacting particles has been verified to a few per cent: from the supernova to the detectors, both photons of light and neutrinos travelled along the same geodesic.

Types of detectors: totally active and sampling

To provide shielding from cosmic ray muons to the level of the background radiation from rock, neutrino telescopes must be deep underground, at a depth of at least 1.5 km water equivalent. On the other hand, they must be in caverns enormous enough to contain a target of sufficiently large mass to record the interactions of the weakly interacting neutrinos or to observe enough nucleons ($\sim 10^{33}$) to hope to see a few decay despite their incredibly long lifetime (over 10^{33} years).

Alternatively, an underground detector can cover a very large area in order to detect the muons produced by neutrino interactions in the surrounding rock. However, the large mass detectors observe both totally contained neutrino interactions depositing a few MeV (those from the sun or supernova) to a few GeV and muons up to a few hundred GeV from neutrinos interacting in the surrounding rock and subsequently traversing the detector.

These detectors fall into two classes: large tanks of water in which the light from particle showers is recorded and passive absorbers in which the shower is sampled by particle trackers and scintillators.

In the water detectors, thousands of photomultipler tubes (simple light-sensitive "cameras") view the water and record the light produced by the remnants of a neutrino interaction or a proton decay. When charged particles travel at speeds faster than the speed of light in the water, they produce the deep blue "Cerenkov" light. These detectors are totally active and are uniquely direction-sensitive, owing to the "bow wave" geometry of the emission of the Cerenkov light.

The first and largest of these detectors, the IMB facility (for Irvine, Michigan, and Brookhaven, the founding institutions) in the United States, deployed a $(20 \text{ m})^3$ tank containing a total of 8 000 tonnes of water with a fiducial mass of 3 300 tonnes. It is being decommissioned and combined with a similar, smaller detector, Kamiokande, at the Kamioka metal mine in Japan to form a larger, second-generation detector, SuperKamiokande (SuperK). The mass of SuperK, 50 000 tonnes total and 22 000 tonnes fiducial, is undoubtedly the largest that can be accommodated by the stability of a rock cave.

The SuperK detector, due to be commissioned in 1996, will extend the search for proton decay to 10^{34} years and refine the study of solar neutrinos. In addition, the current set of detectors indicates a possible lack of muon neutrinos from the decay of particles produced by cosmic rays in the atmosphere. This could be due to a transformation of the neutrinos from one type (muon) to another (electron or tau). In addition to a more sensitive search for these oscillations of atmospheric neutrinos, the SuperK detector is located along a proposed neutrino beam line from the KEK accelerator east of Tokyo. This accelerator beam would allow it to search for neutrino oscillations with a long baseline and a low neutrino energy in the same region as the atmospheric neutrino anomaly.

Due to the large area phototubes that cover 20 per cent of the outer surface of the first detector at the Kamioka mine (Kamiokande), it has superior sensitivity to low energy neutrinos (7 MeV). It has observed neutrinos that point back to the Sun. IMB has had the largest exposure (area × time) to through-going muons from high energy neutrinos, as well as to contained neutrinos. Nevertheless, neither IMB nor Kamiokande has identified point sources of extrasolar neutrinos. Even the new SuperK will be too small, by a factor of 10^3, to see the anticipated sources.

The second type of underground detector achieves large mass in a small volume by using a heavy absorber, usually iron (stone in one case). Charged particle detectors (drift chambers, limited streamer tubes, etc.) are interspersed to reveal the track of a through-going muon or the debris of a neutrino interaction or proton decay. However, a major portion of the interaction is hidden inside the heavy, inactive absorber.

The largest detectors to date (~1 000 tons), in the Frejus tunnel between France and Italy (now decommissioned) and at the Soudan iron mine in the United States, are second-generation devices that have built on the extensive experience of an Italian detector in the Mont Blanc and an Indian detector in the Kolar Goldfields. These tracking detectors, with very different systematic errors from those of the water Cerenkov detectors, give similar results for proton decay and for atmospheric neutrinos, though with lower sensitivity and bigger error bars, owing to their smaller mass and area.

A major new detector of the sampling type, MACRO (for Monopole and Cosmic Ray Observatory), has recently begun to record data. An Italian/American collaboration, it is located in the first national laboratory dedicated to underground physics, the Gran Sasso National Laboratory. It is located in the centre of a highway tunnel in the Apennine mountains east of Rome. Three large halls contain several experiments. The investment in the infrastructure for this laboratory, fully borne by Italy, is some $75 million. The value of the experiments mounted there, in which some ten countries participate, is over $30 million.

MACRO is optimised to present a large area (~1 000 m²) to through-going particles; hence, it will be the most sensitive neutrino telescope for energies larger than 10 GeV. However, it is undoubtedly too small to see the anticipated point sources. In addition to limited streamer tube chambers, the passage of charged particles is registered by ionisation energy, deposited as light, in three planes of liquid scintillator. They, *i.e.* the planes of liquid scintillator, are sensitive to particles travelling much more slowly than the speed of light in the scintillator and therefore depositing much less light than muons.

The low threshold of the scintillator is tailored to detect magnetic monopoles. In addition, the detector is outfitted with sheets of plastic which, when etched, reveal the damage due to the unusually heavy ionisation of a monopole. Further, the signature of a monopole traversing the streamer tubes is also unique. Hence, even a single passage, verified by at least four different detectors, could provide convincing evidence of a monopole. The MACRO detector is unique; it alone can hope to reach the anticipated natural flux of monopoles, the Parker bound. This is the flux that is consistent with the observed magnetic field in our Galaxy.

Another large detector in the Gran Sasso Lab, the LVD (for Large Volume Detector), is a hybrid of the two detector types and was constructed in an Italian/Russian collaboration. Tracking is accomplished by streamer tubes, but the absorbing medium, liquid scintillator, is totally active. The lower energy threshold of the scintillator, and its large volume, could provide unique advantages to this type of detector. It builds upon the experience of two earlier detectors, one in the Mont Blanc tunnel between Switzerland and France, and the other in a cavern in the Baksan Valley in the Caucasus Mountains of Georgia.

Physics requires a detector a thousand times larger

Critical scientific questions remain which are unlikely to be answered by the current generation of underground detectors:

1. Do active galactic nuclei reveal their nature when viewed by neutrino "light"? Can we "see" through the shroud of dust surrounding the centre of the Milky Way (our Galaxy)

and identify a black hole there with neutrinos? For this task, a detector with a surface area of 1 km^2 appears to be necessary, 1 000 times bigger than the largest area detector, MACRO.

2. If supersymmetry (SUSY) is the grand unifying theory, an undiscovered spectrum of particles, inactive since the Big Bang, should exist at higher energies than we have been able to explore to date. They may be caught and stopped in "gravitational wells" like our Sun or the centre of the Earth. There, annihilations of SUSY particles and antiparticles should produce neutrinos of ~10 GeV energy. Unfortunately, the expected rates are too low to be seen in the current generation of underground detectors.

3. Neutrino oscillation searches with great sensitivity require long baselines and low energies to allow enough time for the neutrinos to transform in their rest frame. But long baselines mean negligible rates unless the detectors are much larger in mass than current experiments. Further, controlled experiments with neutrino beams from accelerators (*e.g.* KEK in Japan, CERN in Geneva and Fermilab in Chicago), as well as experiments with neutrinos from the atmosphere, will require detectors with lower systematic errors (*e.g.* multiple detectors and better knowledge of the charged particle flux in the atmosphere).

4. Currently no background is expected for the SuperK proton decay search into the two paradigm decay modes (e$^+$p^0 and nK^0). But since SuperK is only seven times larger than IMB, where there are no candidates, they can at most obtain a handful of events. If the SUSY theory is correct, a megatonne detector is necessary to observe nucleon disintegration.

Prototypes for the "next" detector – underwater or ice

With the large volume of the SuperK detector and the large area of the MACRO detector, the sensitivity of underground detectors has reached its natural limits: rock mechanics will not support larger man-made caverns. Only one route seems promising for the next generation of experiment: Cerenkov detection underwater or in ice.

At depths of three to four km, the background due to charged particles penetrating from cosmic rays is sufficiently attenuated, as is the light from the surface. Two promising locations are being investigated by two different groups: Dumand (for deep underwater muon and neutrino detector) is a collaboration of institutes from Germany, Japan, Switzerland, and the United States that is working off the island of Hawaii. Nestor (for neutrinos from supernovae and TeV sources ocean range) combines expertise from Germany, Greece, Italy, and the United States in a detector off the coast of Greece. Figure A.5 shows an artist's view of the proposed octagonal Dumand array of 216 optical modules. Strings of 24 photomultipliers float vertically up from the bottom. In the Nestor design, Figure A.6, a semi-rigid "tower" provides a support for horizontal clusters of photomultipliers. The shallower depths of a deep lake have already been outfitted by the Lake Baikal collaboration (Germany and Russia) in Siberia. In all these cases, the location has been chosen with a view to minimum currents, an extensive flat bottom close to shore, and minimal biolife.

In ice, four holes 1 km deep have been drilled with hot water in Antarctica and backfilled with strings of photomultipliers and water by the Amanda group (Sweden and the United States). The challenge here is to go deep enough to compress ice bubbles that would scatter and absorb the light. This depth is thought to be sufficient. Beyond it, the cost of drilling rises dramatically with depth.

In these four detectors, the photomultipliers are protected by glass spheres designed to take the hundreds of atmospheres of pressure. Optical fibers are used to get the signals to and from a laboratory on shore for the water experiments.

192

Figure A.5. **Dumand design**

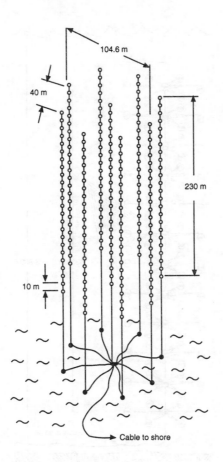

Source: Authors.

The Dumand and Nestor programmes have similar goals:

- to deploy an array of ~200 photomultipliers anchored to the bottom of the ocean at ~4 km depth and rising a few hundred meters toward the surface;
- to observe, at their known rates and angular distributions, through-going muons and atmospheric neutrino interactions with vertices occurring within the array.

The goals of the Baikal and Amanda experiments are also similar, but they must deal with the higher background at their shallow depths.

It is the authors' opinion that each of these efforts should be classified as a prototype for a single focused facility. Each appears to be too small and undermanned and underfunded to realise its ambitious scientific goals. A joint facility that uses the combined strength of all four groups might be a better solution.

Figure A.6. **Nestor design**

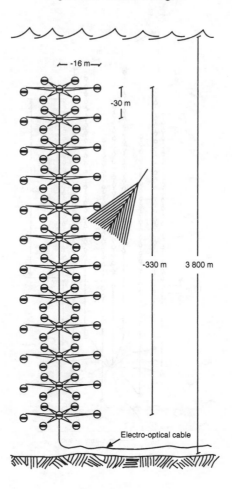

Source: Authors.

Sketch of an international neutrino astroparticle observatory

The outlines of an international underwater facility, drawn from the proposals of Amanda, Baikal, Dumand, and Nestor, based on the experience of IMB and Kamiokande, and extrapolated from SuperK, might consist of up to six nested arrays, staged over a decade or two, each optimised

194

for a different physics goal but overlapped to provide redundancy, cross-correlation, and cross-calibration.

The simplest, least expensive, and therefore the first detector would be a sparce (30 m spacing, determined by light attenuation length), large area (1 km)^2, upward-muon detector. It would search for muon neutrinos from active galactic nuclei, from the centre of the Milky Way and from annihilation of SUSY particles in the Sun and the centre of the Earth. It would be comprised of ten Dumand/Amanda or Nestor type towers, ~100 m high, in order to have sufficient angular (1°) and energy resolution. It could cost up to $100 million.

Experience from the initial array could lead, in a few years, to a similar, smaller detector (300 m)^2 embedded in the first. It would have much better sensitivity in the low energy regime, thanks to a higher density of photomultipliers (spacing of ~5 m, determined by electromagnetic shower size). It would be able to detect electromagnetic showers with a threshold of 5-10 GeV. It could therefore observe electron neutrinos from SUSY annihilation in the Sun and the centre of the Earth. It could measure the ratio of electron to muon atmospheric neutrinos and detect their oscillations. With this energy threshold, electron neutrinos are suppressed relative to muon neutrinos, since the muons that produce the ν_e do not have a chance to decay, making it possible to look for an appearance, which has better signal to noise. This detector would cost perhaps $200 million.

A few years later, a highly instrumented megatonne proton decay detector could be deployed inside the core of the first two arrays, with a sensitivity at the ultimate level achievable on Earth, a few times 10^{35} years for proton decay (ten to 20 times better than SuperK), where the signal is expected to be unresolvable from the atmospheric neutrino background. Because the attenuation length of light would be shorter than the dimensions of the detector, the core would be a volume array, unlike IMB and SuperK which have photomultipliers only on the outer surface. Since solar neutrinos would not be a goal, it would not have the expensive very low (5 MeV) threshold of SuperK. However, the possibility of making such a highly instrumented megatonne detector sensitive to supernovae as far as the Andromeda galaxy deserves further studies. It would give an opportunity to measure neutrino masses, for any flavour, down to about 20 eV, a range that is of cosmological interest. With new large area photomultipliers and extensive R&D on them, the cost of this detector might be as low as $200 million.

This would complete a programme proposed almost 20 years ago by a group that included one of the authors of this annex. Furthermore, this international detector facility could be subsequently upgraded in two possible ways. A ultra-large area array of acoustic detectors (1 000 km^2) could reach neutrino energies in the PeV region. Alternatively, a large area (1 km^2) surface array could serve as a telescope for ultra-high energy γ-rays (> 1 TeV) whose showers penetrate to the surface of the Earth. This facility could be envisaged as the equivalent of CERN for astroparticle physics.

4. Low energy solar neutrino detectors

Solar neutrinos are produced in the deep interior of the Sun through thermonuclear reactions. They are the only probes of the interactions in the core of the Sun. Solar neutrinos range up to 15 MeV in energy and can only be detected by special techniques. They cannot be recorded by high energy neutrino detectors. Owing to their low energy and to our distance from the Sun, solar neutrinos provide a unique opportunity to study neutrino masses and flavour mixing over a wide range that is inaccessible with reactors or with accelerators. Furthermore, this range is of considerable interest in a class of grand unified theories.

Solar neutrino fluxes are presently being measured by four experiments, which are characterised in Table A.4.

Table A.4. **Solar neutrino flux measurements according to four current experiments**

	Threshold	Technique	Starting time	Flux/theory
Kamioka-H$_2$O	7.3 MeV	Cerenkov	1987-	0.5
Homestake-Cl	0.82 MeV	Radiochem	1970-	~0.3
GALLEX-Ga	0.23 MeV	Radiochem	May 1991-	~0.6
SAGE-Ga	0.23 MeV	Radiochem	January 1990-	~0.6

Source: Authors.

The Kamioka experiment is the real-time water Cerenkov experiment described above. The basic process observed is neutrino scattering from electrons which then produce detectable light. The observed neutrino rate is half that predicted by theory.

Since 1967, the pioneering experiment at the Homestake gold mine in South Dakota has been extracting ^{37}Ar atoms created by electron neutrino scattering from ^{37}Cl in a tank containing 615 tonnes of perchloroethylene (cleaning fluid). A production rate of only 0.4 atoms/day is observed, compared to the 1.5/day predicted by the solar models. However, the neutrino energy threshold for ^{37}Ar production is still high, 0.82 MeV, and only solar neutrinos produced in a rare chain of reactions contribute to the rate.

The two gallium-based detectors have the unique advantage of a very low threshold, 0.23 MeV. These experiments extract ^{71}Ge atoms from a 30 tonne Ga target. Gallex (for gallium experiment), a collaboration of mostly French, German, and Italian scientists, is located in the Gran Sasso Laboratory; SAGE (Soviet-America Gallium Experiment) is operated in the Baksan Valley underground laboratory. For both experiments, the observed rate is now around 60 per cent of the expected one.

Thus, all these experiments observe a deficit of solar neutrinos when compared with the expectations raised by detailed solar codes. Unfortunately, the origin of the discrepancies is not yet clear. They could be due to new neutrino properties beyond those of the electroweak standard model, *e.g.* non-zero neutrino masses, and mixing. They could also be due to poorly understood phenomena in the Sun, to problematic input data to the solar models, or to as yet completely unknown causes.

Four experiments are now planned for the near future. Their status is indicated in Table A.5. All of these experiments will bring new insights to bear on the solar neutrino anomaly. For example, SNO (Sudbury Neutrino Observatory) and Icarus could resolve the neutrino composition, v_e versus v_μ plus v_τ. Although none of these experiments, other than Icarus, is costed at above $100 million, their sum total is much higher, and in fact, all measurements should be considered as

Table A.5. **Four solar neutrino experiments for the near future**

Super Kamioka	1996	22 000 tonnes H$_2$O	$60 million	Funded
SNO (Canada)	1996	1 000 tonnes D$_2$O	$50 million	Funded
Borexino (Italy)	~1998	300 tonnes scintillator	$50 million	Proposed
Icarus (Italy)	~2000	2 000 tonnes liquid argon	$100 million	Proposed

Source: Authors.

contributing to the same experiment; only by combining the data from different energy regions and from different neutrino flavour sensitivities will it be possible to obtain full understanding of the solar neutrino spectrum.

Moreover, it is not excluded that, over time, a variation in the solar neutrino flux (either in the high energy or in the low energy range) may occur, revealing recent changes in the rate of thermonuclear fusion reactions in the deep solar interior, or new neutrino properties.

In any case, international co-ordination in staging these experiments appears essential, especially to ensure temporal overlap when decommissioning old experiments and commissioning new ones. The following is clearly a policy question: Should Gallex be stopped in 1997 when the gallium must either be returned or purchased ($\sim$ \$10 million) and the statistical error will be at the level of the systematic error? Like high energy cosmic ray observations, this field would benefit from an international advisory committee.

Conclusions

The field of particle astrophysics is growing. Many experiments on various scales and many new ideas are emerging. Particle astrophysics is now a recognised and highly rated field, which, in the next decade, is likely to attract more than 2 000 particle physicists, if the megaprojects described above are funded along with smaller-scale experiments.

In the absence of any present structure, the authors feel that the OECD, or some other appropriate organisation, should sponsor a representative global advisory committee. This would be a milestone in astroparticle physics, a field which has generally "fallen into the cracks", situated as it is between accelerator physics and astronomy.

As an example in the area of high energy neutrino astrophysics, this committee could facilitate a merger of Amanda, Baikal, Dumand, and Nestor which would result in a single international facility, with resources sufficient for the physics tasks.

However, two caveats must be kept in mind. First, this committee would have the difficult task of comparing projects with very different scientific goals. Second, there is a risk of unnecessary bureaucracy which might delay, rather than expedite, the task of mounting even first-class projects.

Acknowledgements

The authors would like to acknowledge the inspiration in preparing this paper of the following initiators of the field of particle astrophysics: B. Barish, R. Barloutaud, E. Bellotti, V. Berezinsky, A. Chudakov, D. Cline, J. Cronin, T. Damour, M. Davier, R. Davis, A. De Rujula, G. Domogatsky, E. Fiorini, G. Fontaine, S. Glashow, F. Halzen, E. Iarocci, T. Kirsten, T. Koshiba, J. Learned, L. Moscoso, P. Petiau, F. Reines, L. Resvanis, C. Rubbia, J. Stone, A. Suzuki, K.S. Thorne, Y. Totsuka, D. Vignaud, D. Winn, G. Zatsepin, and A. Zichichi.

Acronyms

ADC	Analogue to digital converter
AGS	Alternating Gradient Synchrotron (United States)
ALEPH	Apparatus for LEP Physics
ALICE	A Large Ion Collider Experiment
ANL	Argonne National Laboratory (United States)
APS	American Physical Society
ATF	Accelerator Test Facility
ATLAS	A Toroidal LHC Apparatus (CERN)
BEPC	Beijing Electron-Positron Collider (China)
BNL	Brookhaven National Laboratory (United States)
CAMAC	Computer automated measurement and control
CAT	Computerised axial tomography
CEA	Cornell Electron Accelerator (United States)
CERN	European Laboratory for Particle Physics (Geneva)
CESR	Cornell Electron Storage Ring (United States)
CLIC	CERN Linear Collider
CMS	Compact Muon Solenoid (CERN)
COCOM	Co-ordinating Committee for Multilateral Strategic Export Controls
CP	Product of charge conjugation symmetry and parity symmetry
DELPHI	Detector with Lepton, Photon, and Hadron Identification
DESY	Deutsches Elektronen Synchrotron (Germany)
DORIS	Doppel Ring Speicher (Germany)
ECFA	European Committee for Future Accelerators
EGS	Electron-gamma shower
ELFE	Electron Laboratory for Europe
EMBO	European Molecular Biology Organisation
EPLC	Electron-Positron Linear Collider
EPS	European Physical Society
ESO	European Southern Observatory
ESRF	European Synchrotron Radiation Facility
Fermilab	Fermi National Laboratory (United States), also FNAL
FFTB	Final Focus Test Beam Facility
FSU	Former Soviet Union

G-W-S model	Glashow-Weinberg-Salam model
GUT	Grand Unified Theory
HEPAP	High Energy Physics Advisory Panel (Department of Energy, United States)
HERA	Hadron Elektron Ring Anlage (DESY, Germany)
ICFA	International Committee for Future Accelerators
ICSU	International Council of Scientific Unions
ICHEP	International Conference for High Energy Physics
ISOLDE	Isotope Separation On-line Detector Experiment (CERN)
ISR	Intersecting Storage Ring (CERN)
IUPAP	International Union of Pure and Applied Physics
JINR	Joint Institute for Nuclear Research (Russia)
JLC	Japan Linear Collider
KEK	National Laboratory for High Energy Physics (Japan)
LAMPF	Los Alamos Meson Physics Facility (United States)
LBL	Lawrence Berkeley Laboratory (United States)
LEP	Large Electron-Positron Collider (CERN)
LEP-II	Large Electron-Positron Collider upgrade to 200 GeV (CERN)
LHC	Large Hadron Collider (CERN)
LLNL	Lawrence Livermore National Laboratory (United States)
MOU	Memorandum of Understanding
MRI	Magnetic resonance imaging
NLC	Next linear collider
NMR	Nuclear magnetic resonance
NuPECC	Nuclear Physics European Collaboration Committee
OPAL	Omni-Purpose Apparatus for LEP
PEP	Positron-Electron Project (United States)
PET	Positron emission tomography
PS	Proton Synchrotron
PSI	Paul Scherrer Institute (Switzerland)
QCD	Quantum chromodynamics
QED	Quantum electrodynamics
RF	Radio frequency
RFQ	Radio frequency quadrupole
RHIC	Relativistic Heavy Ion Collider (United States)
SBLC	S-band Linear Collider
SFM	Split field magnet
SLAC	Stanford Linear Accelerator Center (United States)
SLC	Stanford Linear Collider (United States)
SM	Standard Model
SPEAR	Stanford Positron-Electron Accelerator Ring (United States)
SPS	Super Proton Synchrotron (CERN)
SR	Synchrotron Radiation

SSC	Superconducting Super Collider (United States)
SSCL	SSC Laboratory
SUSY	Supersymmetry
TESLA	TeV Energy Superconducting Linear Accelerator
TRISTAN	Transposable Ring Intersecting Storage Accelerator in Nippon (Japan)
TRIUMF	Meson Research Facility (Canada)
UHV	Ultra-high vacuum
UNK	Accelerating and Storage Complex (Russia)
VBA	Very big accelerator
VLEPP	Very Large Electron-Positron Project (Russia)
VME	Versa Module Eurocard

OECD Megascience Forum Expert Meeting on Particle Physics

List of participants

Chairman
Dr. P.A.J. TINDEMANS, Netherlands

Dr. John BOLDEMAN	Australia
Mr. W. REITER	Austria
Mr. Paul LEVAUX	Belgium
Dr. Pardeep AHLUWAHLIA	Canada
Dr. Alan ASTBURY	Canada
Dr. Nigel LLOYD	Canada
Asst. Prof. Peter HANSEN	Denmark
Dr. Jorma HATTULA	Finland
Prof. P. BARUCH	France
Dr. J.-P. REPELLIN	France
Dr. M. CROZON	France
Prof. J. HAÏSSINSKI	France
Prof. M. SPIRO	France
Prof. J.-J. AUBERT	France
Prof. Dr. Björn WIIK	Germany
Prof. Albrecht WAGNER	Germany
Dr. Hermann SCHUNCK	Germany
Prof. Dr. D. WEGENER	Germany
Mr. Axel VOLHARD	Germany
Prof. G. CASTRO	Italy
Prof. Luciano MAIANI	Italy
Prof. Hirotaka SUGAWARA	Japan
Mr. Kazuo WATANABE	Japan
Prof. Koji NAKAI	Japan
Mr. Tsutomu HAMAGUCHI	Japan
Mr. Yukio KAWAUCHI	Japan
Prof. Dr. K.J.F. GAEMERE	Netherlands
Dr. K.H. CHANG	Netherlands
Prof. Jens FEDER	Norway

Asst. Prof. Dr. Tor THORSTEINSEN	Norway
Prof. Enrique FERNANDEZ	Spain
Prof. Görn JARLSKOG	Sweden
Prof. Orjan SKEPPSTEDT	Sweden
Prof. Felicitas PAUSS	Switzerland
Prof. Allan CLARK	Switzerland
Prof. Maurice BOURQUIN	Switzerland
Prof. Hans Christian WALTER	Switzerland
Mr. M. GOTRETT	Switzerland
Dr. Paul E. ZINSLI	Switzerland
Mr. J.-P. RUDER	Switzerland
Mr. R.P. RITZEMA	Switzerland
Prof. D. SAXON	United Kingdom
Dr. Wilmot N. HESS	United Kingdom
Dr. Jerome FRIEDMAN	United States
Dr. Burton RICHTER	United States
Dr. John PEOPLES	United States
Dr. Robert EISENSTEIN	United States and ICFA
Prof. Eugen MERZBACHER	United States
Dr. Edward M. MALLOY	United States
Dr. Harold JAFFE	United States
Dr. Beth ROBINSON	United States

Observers:

Prof. N. KROO	Hungary
Prof. B. SINHA	India
Dr. Augusto GARCIA	Mexico
Prof. A. SKRINSKY	Russia
Prof. N.E. TJURIN	Russia
Ms. A.M. BELOVA	Russia
Mr. V.N. KOLIAKINE	Russia

Invited observers:

Prof. C. LLEWELLYN-SMITH	CERN
Prof. K. HÜBNER	CERN
Dr. L. EVANS	CERN
Dr. W. HOOGLAND	CERN
Prof. Y. YAMAGUCHI	IUPAP
Prof. H. SCHOPPER	UNESCO

Invited speakers:

Prof. W. WILLIS	Nevis Laboratory, Columbia University
Prof. S. WOJCICKI	Stanford University and HEPAP
Prof. G. FLÜGGE	RWTH, Aachen, and ECFA
Prof. A. DONNACHIE	Schuster Laboratory, University of Manchester
Prof. S. YAMADA	Institute of Nuclear Study, Tokyo University

Prof. M. DANILOV   ITEP, Moscow
Dr. L. SULAK   DAPNIA, CEA, Saclay
Prof. E. OTTEN   Institut fur Physik, University of Mainz
Mr. R.P. RITZEMA   United Kingdom

Consultant:
Prof. P. PETIAU   École polytechnique, France

OECD Secretariat

Mr. Nobuo TANAKA   Director, Directorate for Science,
                                    Technology and Industry
Mr. Michael W. OBORNE   Head, Country Studies and Outlook Division
Mrs. Françoise PRADERIE   Sectoral Issues Branch, Megascience Forum
                                    Co-ordinator
Mr. Nobuhiro MUROYA   Consultant
Ms. Sonia GUIRAUD   Assistant

TITLES PUBLISHED IN THE "MEGASCIENCE: THE OECD FORUM" SERIES

MEGASCIENCE AND ITS BACKGROUND
Price: 70 FF in France - 90 FF for mail orders oustside France
ISBN : 92-64-13926-5

* * *

ASTRONOMY
Price: 170 FF in France - 220 FF for mail orders oustside France
ISBN : 92-64-13928-1

* * *

DEEP DRILLING
Price: 170 FF in France - 220 FF for mail orders oustside France
ISBN : 92-64-13956-7

* * *

GLOBAL CHANGE OF PLANET EARTH
Price: 130 FF in France - 170 FF for mail orders oustside France
ISBN : 92-64-14069-7

* * *

OCEANOGRAPHY
Price: 135 FF in France - 175 FF for mail orders oustside France
ISBN : 92-64-14205-3

* * *

NEUTRON BEAMS AND SYNCHROTRON RADIATION SOURCES
Price: 185 FF in France - 240 FF for mail orders oustside France
ISBN : 92-64-14249-5

* * *

ERRATUM

PARTICLE PHYSICS

(92 95 01 1) ISBN 92-64-14329-7

Please note that, in Figures 3, 4, 5 and 7 (pages 44, 45 and 50), the neutrino is incorrectly indicated by (n) and should appear as (ν).

MAIN SALES OUTLETS OF OECD PUBLICATIONS
PRINCIPAUX POINTS DE VENTE DES PUBLICATIONS DE L'OCDE

ARGENTINA – ARGENTINE
Carlos Hirsch S.R.L.
Galería Güemes, Florida 165, 4° Piso
1333 Buenos Aires   Tel. (1) 331.1787 y 331.2391
Telefax: (1) 331.1787

AUSTRALIA – AUSTRALIE
D.A. Information Services
648 Whitehorse Road, P.O.B 163
Mitcham, Victoria 3132   Tel. (03) 873.4411
Telefax: (03) 873.5679

AUSTRIA – AUTRICHE
Gerold & Co.
Graben 31
Wien I   Tel. (0222) 533.50.14

BELGIUM – BELGIQUE
Jean De Lannoy
Avenue du Roi 202
B-1060 Bruxelles   Tel. (02) 538.51.69/538.08.41
Telefax: (02) 538.08.41

CANADA
Renouf Publishing Company Ltd.
1294 Algoma Road
Ottawa, ON K1B 3W8   Tel. (613) 741.4333
Telefax: (613) 741.5439

Stores:
61 Sparks Street
Ottawa, ON K1P 5R1   Tel. (613) 238.8985
211 Yonge Street
Toronto, ON M5B 1M4   Tel. (416) 363.3171
Telefax: (416)363.59.63

Les Éditions La Liberté Inc.
3020 Chemin Sainte-Foy
Sainte-Foy, PQ G1X 3V6   Tel. (418) 658.3763
Telefax: (418) 658.3763

Federal Publications Inc.
165 University Avenue, Suite 701
Toronto, ON M5H 3B8   Tel. (416) 860.1611
Telefax: (416) 860.1608

Les Publications Fédérales
1185 Université
Montréal, QC H3B 3A7   Tel. (514) 954.1633
Telefax : (514) 954.1635

CHINA – CHINE
China National Publications Import
Export Corporation (CNPIEC)
16 Gongti E. Road, Chaoyang District
P.O. Box 88 or 50
Beijing 100704 PR   Tel. (01) 506.6688
Telefax: (01) 506.3101

DENMARK – DANEMARK
Munksgaard Book and Subscription Service
35, Nørre Søgade, P.O. Box 2148
DK-1016 København K   Tel. (33) 12.85.70
Telefax: (33) 12.93.87

FINLAND – FINLANDE
Akateeminen Kirjakauppa
Keskuskatu 1, P.O. Box 128
00100 Helsinki
Subscription Services/Agence d'abonnements :
P.O. Box 23
00371 Helsinki   Tel. (358 0) 12141
Telefax: (358 0) 121.4450

FRANCE
OECD/OCDE
Mail Orders/Commandes par correspondance:
2, rue André-Pascal
75775 Paris Cedex 16   Tel. (33-1) 45.24.82.00
Telefax: (33-1) 49.10.42.76
Telex: 640048 OCDE
Orders via Minitel, France only/
Commandes par Minitel, France exclusivement :
36 15 OCDE
OECD Bookshop/Librairie de l'OCDE :
33, rue Octave-Feuillet
75016 Paris   Tel. (33-1) 45.24.81.67
(33-1) 45.24.81.81
Documentation Française
29, quai Voltaire
75007 Paris   Tel. 40.15.70.00
Gibert Jeune (Droit-Économie)
6, place Saint-Michel
75006 Paris   Tel. 43.25.91.19
Librairie du Commerce International
10, avenue d'Iéna
75016 Paris   Tel. 40.73.34.60
Librairie Dunod
Université Paris-Dauphine
Place du Maréchal de Lattre de Tassigny
75016 Paris   Tel. (1) 44.05.40.13
Librairie Lavoisier
11, rue Lavoisier
75008 Paris   Tel. 42.65.39.95
Librairie L.G.D.J. - Montchrestien
20, rue Soufflot
75005 Paris   Tel. 46.33.89.85
Librairie des Sciences Politiques
30, rue Saint-Guillaume
75007 Paris   Tel. 45.48.36.02
P.U.F.
49, boulevard Saint-Michel
75005 Paris   Tel. 43.25.83.40
Librairie de l'Université
12a, rue Nazareth
13100 Aix-en-Provence   Tel. (16) 42.26.18.08
Documentation Française
165, rue Garibaldi
69003 Lyon   Tel. (16) 78.63.32.23
Librairie Decitre
29, place Bellecour
69002 Lyon   Tel. (16) 72.40.54.54

GERMANY – ALLEMAGNE
OECD Publications and Information Centre
August-Bebel-Allee 6
D-53175 Bonn   Tel. (0228) 959.120
Telefax: (0228) 959.12.17

GREECE – GRÈCE
Librairie Kauffmann
Mavrokordatou 9
106 78 Athens   Tel. (01) 32.55.321
Telefax: (01) 36.33.967

HONG-KONG
Swindon Book Co. Ltd.
13–15 Lock Road
Kowloon, Hong Kong   Tel. 366.80.31
Telefax: 739.49.75

HUNGARY – HONGRIE
Euro Info Service
Margitsziget, Európa Ház
1138 Budapest   Tel. (1) 111.62.16
Telefax : (1) 111.60.61

ICELAND – ISLANDE
Mál Mog Menning
Laugavegi 18, Pósthólf 392
121 Reykjavik   Tel. 162.35.23

INDIA – INDE
Oxford Book and Stationery Co.
Scindia House
New Delhi 110001   Tel.(11) 331.5896/5308
Telefax: (11) 332.5993
17 Park Street
Calcutta 700016   Tel. 240832

INDONESIA – INDONÉSIE
Pdii-Lipi
P.O. Box 269/JKSMG/88
Jakarta 12790   Tel. 583467
Telex: 62 875

ISRAEL
Praedicta
5 Shatner Street
P.O. Box 34030
Jerusalem 91430   Tel. (2) 52.84.90/1/2
Telefax: (2) 52.84.93
R.O.Y.
P.O. Box 13056
Tel Aviv 61130   Tél. (3) 49.61.08
Telefax (3) 544.60.39

ITALY – ITALIE
Libreria Commissionaria Sansoni
Via Duca di Calabria 1/1
50125 Firenze   Tel. (055) 64.54.15
Telefax: (055) 64.12.57
Via Bartolini 29
20155 Milano   Tel. (02) 36.50.83
Editrice e Libreria Herder
Piazza Montecitorio 120
00186 Roma   Tel. 679.46.28
Telefax: 678.47.51
Libreria Hoepli
Via Hoepli 5
20121 Milano   Tel. (02) 86.54.46
Telefax: (02) 805.28.86
Libreria Scientifica
Dott. Lucio de Biasio 'Aeiou'
Via Coronelli, 6
20146 Milano   Tel. (02) 48.95.45.52
Telefax: (02) 48.95.45.48

JAPAN – JAPON
OECD Publications and Information Centre
Landic Akasaka Building
2-3-4 Akasaka, Minato-ku
Tokyo 107   Tel. (81.3) 3586.2016
Telefax: (81.3) 3584.7929

KOREA – CORÉE
Kyobo Book Centre Co. Ltd.
P.O. Box 1658, Kwang Hwa Moon
Seoul   Tel. 730.78.91
Telefax: 735.00.30

MALAYSIA – MALAISIE
Co-operative Bookshop Ltd.
University of Malaya
P.O. Box 1127, Jalan Pantai Baru
59700 Kuala Lumpur
Malaysia   Tel. 756.5000/756.5425
Telefax: 757.3661

MEXICO – MEXIQUE
Revistas y Periodicos Internacionales S.A. de C.V.
Florencia 57 - 1004
Mexico, D.F. 06600   Tel. 207.81.00
Telefax : 208.39.79

NETHERLANDS – PAYS-BAS
SDU Uitgeverij Plantijnstraat
Externe Fondsen
Postbus 20014
2500 EA's-Gravenhage   Tel. (070) 37.89.880
Voor bestellingen: Telefax: (070) 34.75.778

OECD PUBLICATIONS, 2 rue André-Pascal, 75775 PARIS CEDEX 16
PRINTED IN FRANCE
(92 95 01 1) ISBN 92-64-14329-7 – No. 47647 1995